Green Energy Futures

DOI: 10.1057/9781137584434.0001

Also by David Elliott

THE CONTROL OF TECHNOLOGY (*with R. Elliott*)

MAN MADE FUTURES (*edited with N. Cross and R. Roy*)

THE POLITICS OF TECHNOLOGY (*edited with G. Boyle and R. Roy*)

THE POLITICS OF NUCLEAR POWER (*with P. Coyne, M. George and P. Lewis*)

THE LUCAS PLAN (*with H. Wainwright*)

ENTERPRISING INNOVATION (*with V. Mole*)

PRIVATISING ELECTRICITY (*with J. Roberts and T. Houghton*)

ENERGY, SOCIETY AND ENVIRONMENT

A SOLAR WORLD: Climate Change and the Green Energy Revolution

NUCLEAR OR NOT?

FUKUSHIMA: Impacts and Implications

SUSTAINABLE ENERGY

RENEWABLES: A Review of Sustainable Energy Supply Options

DOI: 10.1057/9781137584434.0001

palgrave▸pivot

Green Energy Futures: A Big Change for the Good

David Elliott

Emeritus Professor of Technology Policy

The Open University, UK

palgrave
macmillan

DOI: 10.1057/9781137584434.0001

First published 2015 by
PALGRAVE MACMILLAN

Palgrave Macmillan in the UK is an imprint of Macmillan Publishers Limited, registered in England, company number 785998, of Houndmills, Basingstoke, Hampshire RG21 6XS.

Palgrave Macmillan in the US is a division of St Martin's Press LLC, 175 Fifth Avenue, New York, NY 10010.

Palgrave Macmillan is the global academic imprint of the above companies and has companies and representatives throughout the world.

Palgrave® and Macmillan® are registered trademarks in the United States, the United Kingdom, Europe and other countries.

ISBN: 978–1–137–58444–1 EPUB
ISBN: 978–1–137–58443–4 PDF
ISBN: 978–1–137–58442–7 Hardback

A catalogue record for this book is available from the British Library.

A catalog record for this book is available from the Library of Congress.

www.palgrave.com/pivot

DOI: 10.1057/9781137584434

Contents

DOI: 10.1057/9781137584434.0001

Preface

This book is based on the assumption that the use of fossil fuels has to be halted, primarily to avoid the economic, social and environmental impacts of climate change. With the divestment movement catching on and fossil companies becoming worried about the economic viability of their fossil investment as governments tighten up emission regulations, change is underway. But what will replace fossil fuel?

This book looks at whether there is a way forward using renewable energy sources and energy efficiency initiatives to cut emissions from fossil fuels while avoiding nuclear power. It concludes that nuclear is unlikely to have much of a role in future, and argues that the pro- and anti-nuclear debate has absorbed too much time and energy over the years, to the detriment of the more relevant, interesting and increasingly urgent debate over what sort of sustainable 'green energy' renewable/efficiency mix we need. That is the focus of this book, which explores the problems and implications of shifting to greener, cleaner energy sources and the policy changes that are underway. It argues there is no one green future. There is a range of possible options of various types and scales: we need to choose amongst them. And this book offers an overview of the key technical, economic and environmental issues to aid that process.

DOI: 10.1057/9781137584434.0002

Quick Guide to Energy Units and Energy Terms

The **Power** rating of an energy conversion device is usually measured in **watts** and multiples – kilowatts (1,000 watts), megawatts (1,000 kW), gigawatts (1,000 MW), terawatts (1,000 GW). A typical small kettle will have a power rating of around 1 kW, a large power plant 1 GW.

The amount of **energy** an energy conversion device supplies or uses is measured in 'power times time', typically kilowatt-hours (kWh), the usual unit for bills, and multiples, for example, MWh. So a 1 kW rated electric fire running for an hour would convert 1 kWh of electrical energy into heat (with losses).

As these definitions should make clear, the common usage of the terms 'power' and 'energy' as interchangeable is incorrect and unhelpful, though hard to avoid. The same goes for the widespread use of 'power' as a shorthand for 'electricity'. Moreover, strictly speaking, energy cannot be generated, or indeed 'used'; it can only be converted, with losses, from one form to another.

The term 'Green energy' is a shorthand for the various types of low-carbon energy systems which use energy sources that are naturally renewed and are therefore sustainable in the future.

Common abbreviations used in this text are as follows:

CCS – Carbon Capture and Storage (carbon dioxide gas sequestration)

CfD – Contracts for Difference (a new UK support scheme based on contract auctions)

DOI: 10.1057/9781137584434.0003

CHP – Combined Heat and Power (co-generation of heat and power)
CSP – Concentrating Solar Power (focused solar systems for electricity production)
EROEI – Energy Return on Energy Invested (a measure of energy input to output)
FiTs – Feed in Tariffs (a support scheme widely used in the EU)
PV – photovoltaic solar (solar-electric cells)
RO – Renewable Obligation (a soon-to-be-retired UK support scheme)

DOI: 10.1057/9781137584434.0003

1

Introduction: What Are the Options?

Abstract: *Climate change, along with falling air quality, are key environmental concerns and seem to be related to human activities, most obviously the combustion of increasing amounts of fossil fuel. There are several options for reducing or avoiding these problems. A shift to using nuclear energy is seen by some as one, but there are significant technical, economic, safety and security problems. Reducing energy waste is a more likely contender, but there are limits, and however much energy efficiency is increased, there will still be a need for energy supplies. That leaves renewable energy sources as a key hope. This book asks, what are the problems and can these new green sources be used to deliver energy reliably and economically on a significant scale?*

Keywords: climate change; energy saving; nuclear power; renewable energy

Elliott, David. *Green Energy Futures: A Big Change for the Good*. Basingstoke: Palgrave Macmillan, 2015.
DOI: 10.1057/9781137584434.0004.

1.1 A big change is needed

Access to energy services is vital to modern life, but there are growing concerns about whether the current range of energy sources can be used into the future. Some worry that fossil fuel reserves will be exhausted, and there are debates about when 'peak oil', the point at which use outstrips production, will occur. Some say it already has, others say that, with shale oil and other finds, there are still decades in hand, but the debate is just about when, not if, oil will become scarce (Brandt et al., 2013). Similarly, though later, for gas (including shale gas) and later still for 'peak coal' (Maggio and Cacciola, 2012).

However, the reality seems to be that not much of whatever there is left of these fossil energy sources can be burnt off without risking what the Intergovernmental Panel on Climate Change warns could be 'severe, widespread, and irreversible impacts' from climate change (IPCC, 2014). In parallel there are increasing impacts on air quality, this reaching crisis point in some newly industrialising countries, most visibly in China.

This book is based on the assumption that the use of fossil fuels has to be halted, probably long before this is forced on us by the inevitable ultimate depletion of these resources. As with the resources estimates, there are debates on timescale, and for example, over how serious climate change related impacts might be and on how quickly, and where, they will occur, but few deny that it is a major and increasingly urgent problem, and even fewer deny that air pollution is having major heath impacts.

There may be ways to limit some of these impacts, or even adapt to them, but in longer term we have to deal with the problems at source and stop burning fossil fuels. That will involve a major change. Over 80% of the energy used globally comes from these sources – coal, oil and gas – used for heating, electricity production and to power vehicles. The aim of this book is to ask whether their use can be phased out, and if so, what will be the problems and implications of shifting to greener, cleaner renewable energy sources.

1.2 Climate change

Ever since the industrial revolution, the combustion of fossil fuel has expanded, coal at first, then oil and more recently (natural) gas, all extracted from underground strata. Burning these fuels releases carbon

DOI: 10.1057/9781137584434.0004

dioxide gas into the atmosphere and it has an impact: it increases the so-called greenhouse effect. As with the panes of glass in a garden greenhouse, some solar radiation that would otherwise not be retained is trapped by a range of gases in the upper atmosphere, so the incident solar radiation heats up the greenhouse, in this case the earth. There is a natural balance in this process which keeps the planet's basic temperature at habitable levels. However, adding extra carbon dioxide, along with other greenhouse gasses such as methane (mostly from increased intensive agriculture), disturbs this balance, so that the average global temperature begins to rise: that's global warming.

So far that is incontrovertible. It is widely agreed that temperatures have risen and also that this has had an effect on the climate system. In a joint report, the United Kingdom's Royal Society and the US National Academy of Sciences said: 'It is now more certain than ever, based on many lines of evidence, that humans are changing Earth's climate. The atmosphere and oceans have warmed, accompanied by sea-level rise, a strong decline in Arctic sea ice, and other climate-related changes. The evidence is clear' (Royal Society/NAC, 2014).

Certainly, the vast majority of scientists are said to think that climate change, due to human activities, is real, up to 97% in some surveys (Cook et al., 2013), and although there are claims that the extent of consensus is less (Tol, 2014), there seems to be wide agreement that climate change will have significant impacts, with the Royal Society/NACS claiming that 'further climate change is inevitable; if emissions of greenhouse gases continue unabated, future changes will substantially exceed those that have occurred so far.' That means, some say, that about 80% of the coal in the ground will have to stay unused (McGlade and Ekins, 2015).

However, there are still debates on how much temperatures will rise in future and what the longer-term impacts will be, driven in part by skeptical minority views, often amplified by the media's inevitable focus on conflict rather than consensus. Some 'contrarians' claim that most of the temperature rise has been due to other, natural, causes, and some say the rises will tail off. They dispute the models that have been produced. The fact that recent temperature rises have been lower than many expected gives them some ammunition, although the climate modelers have come up with various explanations for this. For example, some suggest that the heat has been absorbed in the depths of the oceans, but there is much debate about causes and effects, with some saying that the temperature pause may last for 10–20 years.

DOI: 10.1057/9781137584434.0004

Uncertainty and doubt make it hard to come to an agreement regarding measures for responding to climate change. If it will only be mild and slow, as some contrarians argue, then simple *adaptation* measures should be sufficient, that is, coping with *impacts*. That argument is popular since it avoids having to adopt more radical and potentially costly and disruptive *mitigation* approaches, dealing with the *cause*, chiefly by phasing out fossil fuel use.

Some contrarians claim that global warming will have benefits. That seems to be a line taken by the UK-based Global Warming Policy Foundation, which also says that, if it turns out to be real, negative and significant, it can best be dealt with by adaptation rather than what they see as expensive mitigation measures. Its report on sea-level changes said: 'It is the height of folly, and waste of money, to attempt to "control" the size or frequency of damaging natural events by expecting that reductions in human carbon dioxide emissions will moderate climate "favourably", whether that be putatively sought from a moderation in the frequency and intensity of damaging natural events or by a reduction in the rate of global average sea-level rise' (de Lange and Carter, 2014).

Some contrarians argue that even if it is real, climate change is not that important compared with other global environmental and health issues. Danish contrarian Bjorn Lomborg says that 'Global warming pales when compared to many other global problems. While the WHO estimates 250,000 annual deaths from global warming in 30 years, 4.3 million die right now each year from indoor air pollution, 800 million are starving, and 2.5 billion live in poverty and lack clean water and sanitation'. And anyway, he adds, our approach to dealing with climate change is wrong. Like the Global Warming Policy Foundation, he is doubtful about the value of renewable energy as a response (Lomborg, 2014). It could of course be countered that these issues are not separate and independent. A focus on renewables, in response to climate change, could help address some of the issues Lomborg raises, for example, laying the basis for local economic growth and reducing the need to use polluting firewood and dung.

While debates on policy are fair enough, some think the endless debate on climate science is unhelpful and often of low quality (Dana, 2014). It is important to continually challenge assumptions and check data: debate and conflicts are the lifeblood of science, which moves through periods of doubt and then consensus. But at some point a halt has to be called, for example, it is now clear that the world is not flat and that it orbits the sun. That level of certainty may not have been attained yet over climate

DOI: 10.1057/9781137584434.0004

change, but there are strong indications that there are growing problems so that urgent action is needed. That view is also backed by increasing numbers of the public. In a Populus UK public opinion poll in 2014, 73% wanted world leaders to agree to a global climate deal and 66% thought action must take place now and only 20% felt it could wait a few years (DECC, 2014).

1.3 What should be done?

While concern in many countries is high, the barrage of contrarian views, as relayed by the media, and uncertainties about what might be done seem to have had an impact on some others, for example, in the United States and Australia, where global warming and responses to it are very politicised issues, with many people expressing disbelief. Given that both countries have experienced many very severe weather-related shocks in recent years, this may be surprising, but it remains the case that no one weather event can necessarily be directly liked to climate change. However, it is also true that in both countries, as globally, fossil fuel interests are very powerful, although, around the world, they are increasingly being challenged as awareness of, and concern about, climate change grows (Berners-Lee and Clark, 2013; Klein, 2014).

Despite the uncertainties, governments around the world, to varying degrees, have developed policies for reducing emissions and impacts, both nationally and via international agreements (Marquina, 2010; Dupont and Oberthür, 2015). However, there are disagreements about response strategies, and crucially, at the global level, about who should pay. It seems clear that, whatever happens, adaptation measures will be needed to deal with increased flooding, storms, droughts, wild fires and heat waves. That will hit some countries hard.

Many poorer countries may not be able to cope financially and will need external help from rich countries. Moreover, the longer-term solution of moving away from fossil fuel may not appear to be economically viable for them, and in any case will not help them deal with the impacts in the short term. And yet, it can be argued, unless all countries start making this transition we may all be doomed – if you believe the modeling.

There has been something of a political stand-off. The poor developing countries often claim that the rich countries, who have benefited in

DOI: 10.1057/9781137584434.0004

the past from burning fossil fuels, should shoulder some, or even most, of the cost, for example, by contributing to aid programmes. There is, however, now some movement on this issue. Agreement in principle has been reached on establishing a $100 billion p.a. Green Climate Fund by 2020, with donations from the major industrial countries: the United States recently provided $3 billion.

For the longer term, policy debates and attempted negotiations continue at the annual global Conference of Parties (COPs) to the UN Framework Convention on Climate Change, although these days expectations that the COPs will lead to much of significance are low. The global accord reached at the Kyoto climate summit in 1997 (averaging a 5.2% global emission cut) only ran up to 2012. A loose commitment to a follow-up Kyoto II protocol has been thrashed out, but it is not legally binding. The stumbling blocks have chiefly been the United States and China, which have seemed happy enough to commit to ramping up green energy technology (in competition with each other), but did not want to accept binding constraints on emissions. However, with air quality an urgent issue in China, emission limits there now seem likely (a commitment has been made to halt the rise by around 2030), and the United States, under Obama, evidently is now serious about reducing emissions from coal (by 30% by 2030) and overall by 26–28% by 2025. So the COPs might be a bit more productive.

There are of course deviants, like Japan, which after the Fukushima nuclear disaster, reneged on its emission reduction targets, and Australia, which has experienced if anything even more severe extremes of weather recently than the United States, but is heading off in the opposition direction – cutting just about all its climate policies and initiatives. The EU remains on message and is making progress on its renewable and climate targets (a 20% emission cut by 2020 and possibly 40% by 2030), but is constrained from going further and faster both economically and politically, by the leftovers of the recession and the swing to the political right in many EU countries. Russia remains an anomaly, focused on its huge fossil exports, for example, to Eastern European countries.

The rest of the world? Understandably, as noted earlier, most developing countries want help from rich industrial nations to meet the cost of limiting emissions and dealing with impacts. Unsurprisingly, aid of that sort is something that has been agreed only in rather broad terms.

As things stand at present, the programmes and policies that are in place may not hold average global temperatures below the 2° C rise that

DOI: 10.1057/9781137584434.0004

many think is a crucial threshold, although there are some hopeful signs. The International Energy Agency (IEA) has claimed that emissions from the energy sector in 2014 were at the same level as in 2013, the first time a leveling off or reduction has occurred outside a recession in 40 years, which suggests that policy responses, rather than economic factors, led to zero growth in emissions (Briggs, 2015).

There is also some movement from the fossil fuel interests. With concerns about climate change growing, and, more directly, worries being expressed that fossil assets would become worthless as governments adopted tighter emission regulations, the big oil companies have been beginning to react. Although there may be unethical investors ready to fill the gap, 'divestment' initiatives may also be having an impact, with pension funds, charitable agencies and universities withdrawing their investments in fossil fuel. So too may the potential threat of legal action by climate change victims to reclaim damages. One report, in effect adopting a naming and shaming approach, listed 90 global companies which it said produced 63% of the cumulative global emissions of industrial carbon dioxide and methane between 1751 and 2010. They include big coal and oil companies (Heade, 2014). This sort of campaigning seems to be having an impact. For example, under pressure from shareholders, Shell says it will play a more active role in responding to climate change and warns other companies that they should too (Shell, 2014).

In its 2015 *Annual Energy Outlook*, BP accepted that carbon dioxide emission levels from burning fossil fuels are unsustainable. However, BP warned that renewable energy sources will struggle to keep pace with growing demand for energy, especially for power in Asia: 'The rapid growth of renewables currently depends on policy support in most markets, as renewables tend to be more expensive than coal or gas-fired power. As renewables grow in volume, the burden of this policy support can become a constraint on growth. To maintain rapid growth, the costs of renewable power need to keep falling, reducing the subsidy required per unit of power' (BP, 2015).

So, for good or ill, they see fossil fuels still booming into the future. So do the IEA and the World Energy Council (WEC), although, as we shall see later, they both have scenarios in which renewables ramp up much faster, while IRENA, the International Renewable Energy Agency, claims that, as markets build and the technology improves, renewable energy costs are falling and are also offset by fuel and health cost savings from using less fossil fuel (IRENA, 2014).

DOI: 10.1057/9781137584434.0004

Certainly, renewables are expanding. BP sees them pushing nuclear out, although, unsurprisingly, the nuclear energy lobby sees it all differently. Who is right? Is nuclear the answer? Or is there a way forward that cuts emissions from fossil fuels and avoids nuclear power? Could a switch be made to using renewable energy source as the main the way ahead? Or must frugal lifestyles be adopted? This book aims to explore these issues and to look at the key choices and options available, focusing first on the technological options, and later moving on to options for social change.

1.4 In praise of nuclear

Nuclear power is a much-touted technological option, and the most developed non-fossil 'baseload' energy source, hydro apart. Nuclear power plants do not generate carbon emissions directly, so nuclear power is an obvious option for responding to climate change. There are powerful lobbying organisations and companies backing it, as well as some opposing it, along with plenty of web sites, though surprisingly few recent books supporting it. But in *Why We Need Nuclear Power: The Environmental Case,* published by Oxford University Press, retired US radiation biologist Michael H. Fox argues that nuclear power is essential to slow down the impact of global warming. Although he accepts that wind and solar can contribute, he says we need a 'reliable' source to meet large-scale energy demands and break our dependence on fossil fuels and claims that nuclear power is the best solution to our environmental crisis. In a Blog 'taster', he says renewables can supply only 20% of electricity at most (Fox, 2014).

Adopting a land-use and biodiversity approach, in a paper in the journal *Conservation Biology* entitled 'Key Role for Nuclear Energy in Global Biodiversity Conservation', Prof. Barry Brook and Prof. Corey Bradshaw come to similar conclusions. They say 'for many countries-including most high energy-consuming nations in East Asia and Western Europe with little spare land and already high population densities - the options for massive expansion of renewable energy alternatives are heavily constrained'. Instead, by contrast to the alleged high land-use and biodiversity-loss impact from renewables like wind and solar (a contention that will be revisited in Chapter 2 later), they claim that 'based on an objective and transparent analysis of our sustainable energy choices,

DOI: 10.1057/9781137584434.0004

we have come to the evidence-based conclusion that nuclear energy is a good option for biodiversity conservation (and society in general)'.

That is also the main conclusions of an Open Letter to environmentalists that they fronted, although that is a little more circumspect. While the paper sees nuclear as offering 'prospects for being a principal cure for our fossil-fuel addiction', the Open Letter portrays nuclear as playing a role 'as part of a range of sustainable energy technologies that also includes appropriate use of renewables, energy storage and energy efficiency' (Brook and Bradshaw, 2014a).

Although the paper focuses on the merits of nuclear, and especially the idea of Integral Fast Reactors perhaps using thorium, it says 'making a case for a major role for nuclear fission in a future sustainable energy mix does not mean arguing against energy efficiency and renewable options. Under the right circumstances, these alternatives might also make important contributions' (Brook and Bradshaw, 2014b).

Given that by 2013 renewables were supplying 22% of global electricity compared to nuclear's 11%, that is an understandable caveat (REN21, 2014). But what of the future? As we shall see, many think that renewables could and should expand rapidly, but could nuclear also expand to make a major contribution? There has been talk of a nuclear renaissance, driven at least in part by the need to respond to climate change. Is a major expansion of nuclear power likely or realistic?

1.5 Nuclear limits

The 2015 technology roadmap for nuclear energy, published by the Nuclear Energy Agency and the IEA, suggests that nuclear capacity needs to more than double, to around 930 giga watts (GW), by 2050, to help limit global warming to 2° C (IEA, 2015). An earlier IEA nuclear roadmap, in 2010, put the target 2050 nuclear capacity higher, at 1,200 GW. But even if the IEA's very ambitious target for nuclear was achieved, it would only cut emissions by 2.5 gigatonnes of CO_2 per year, against current annual global emissions from all sources of around 50 GT (Evans, 2015). So the nuclear contribution would be relatively small, and that is ignoring the likely increased emissions from fuel production as high-grade uranium ore became scarcer.

Is an expansion on that scale likely? Not on current progress. The International Atomic Energy Agency says 'the share of nuclear power

DOI: 10.1057/9781137584434.0004

in total global electricity generation decreased for the tenth year in a row, to less than 11% in 2013, the lowest value since 1982' (IAEA, 2014). Moreover, the independent World Nuclear Industry Status Report 2014 did not see it getting better anytime soon (WNISR, 2014). It noted that at least 49 of the total 69 new reactor construction projects, including 75% of China's projects, have encountered delays, of 18–30 months in the case of China's AP1000 reactors, and of 13–15 months with its two European Pressurised-water Reactors, and in the EU (with the French and Finnish EPRs) of several years. Phase-outs continued around the world and construction of several new plants has been delayed, in some case some indefinitely, while some have been abandoned entirely.

It is perhaps not surprising then that Steve Kidd, one time leading nuclear lobbyist with the World Nuclear Association, has had a rethink. In an article in *Nuclear Engineering International* he says 'we have seen no nuclear renaissance' and he outlines his new view. He says 'the high and rising nuclear share in climate-friendly scenarios is false hope, with little in the real outlook giving them any substance'.

Does this mean he has given up the nuclear dream? No, but he says the nuclear lobby has to 'abandon climate change as a prime argument for supporting a much higher use of nuclear power to satisfy rapidly rising world power needs'. That seems partly since he fears that 'there is a significant risk in nuclear hitching itself to this type of view, as it may eventually be found to be unproven and in that case the nuclear industry, along with the renewables sector, will be discredited'. But perhaps also since it had in any case already been a problem: 'The nuclear industry giving credence to climate change from fossil fuels has simply led to a stronger renewables industry', whereas 'nuclear seems to be "too difficult" and gets sidelined'. Instead he argues that the other alleged benefits of nuclear should be the focus. So his revised strategy is one of selling nuclear 'on grounds of cheapness, reliability and security of supply' (Kidd, 2015).

Is this a realistic approach? Leaving aside Fukushima, in the United Kingdom, the past few years have seen old nuclear plants suffering sudden unexpected temporary shutdowns due to faults, and in the United States several old plants have been permanently closed early due to technical economic problems. They could no longer compete with cheaper alternatives. Belgian nuclear plants have been closed after thousands of cracks were found in the pressure vessels. Reliability thus does not seem a strong argument, although it is claimed that new plants will be better, and cheaper. That claim is also hard to accept, given the dramatic escalation of the cost of the two European Pressurised-Water Reactors being built in the EU, due to continuing construction delays and other problems. Delays and cost over-runs have also been experienced with the few new plants being built in the United States.

It is conceivable that new technologies will eventually emerge which will do better and also address some of the other problems with nuclear power, for example, reliance on ever-decreasing reserves of high-grade uranium, the production of radioactive wastes which have to be kept safe somewhere for many thousands of year, and security risks associated with a technology which can be used to make nuclear weapons (Elliott, 2010).

Some countries are still pressing ahead with nuclear programmes at various levels and with varying degrees of success (notably the United Kingdom, United States, China, India, South Korea and Russia), but many others have not gone down that route (including, in the EU, Austria, Denmark, Greece, Ireland, Portugal), while, after the Fukushima nuclear disaster, Japan closed all its plants, and although some may restart, there is little chance of new plants being built there. After Fukushima, opposition to nuclear grew around the world, and governments were forced by public opinion to abandon their nuclear plans (e.g., 94% in a referendum in Italy voted against nuclear), with some phasing out existing plants (e.g., Germany, Belgium, Switzerland) or cutting back radically (France) (Elliott, 2013a).

Some still look to a new generation of breeder reactors, possibly using thorium, or even to nuclear fusion, as the long-term hope. However, at best, these are all decades away, at any significant scale, and have their own problems, as well as unknown costs. Nuclear may still be with us, making a relatively minor contribution, but for the moment, and for a while ahead, it seems we must look elsewhere for an effective response to environmental problems like climate change and air pollution due to burning fossil fuels.

1.6 Save it: energy efficiency

The obvious way to reduce emissions is to avoid generating energy, or at least to use it more efficiently. The potential for reducing energy waste is huge. The EU is aiming to reduce overall energy use by 27% by 2030, while

DOI: 10.1057/9781137584434.0004

Germany and France have set targets of reducing energy use by 50% by 2050. This will not be easy. However, the industrial countries have been profligate in their use of energy, since it was relatively cheap. So the good news is that there are many easy and cheap options for making energy and cash savings, although once they have been exhausted it will get more expensive; there is only so much 'low hanging fruit'. Even so, with some getting cheaper with mass production, new, more energy efficient technologies can help to reduce energy use. So can lifestyle and behavioural change, although this may be painful. While some look to 'nudge' measures, others see draconian personal carbon rationing as the way ahead.

Overall, domestic electricity use in the United Kingdom has fallen by 25% since 2005. And the UK government has estimated that the right energy efficiency framework could save the equivalent of the output of 22 new power stations by 2020. That should be a key priority. But even if large cuts in energy use can be made, there will still be a need for energy supply. If emissions are to be cut, where will it come from?

There are ways to generate energy using fossil fuels with lower emissions, for example, by capturing and storing them in underground rock strata in empty oil and gas well-sites. That is expensive, with uncertain impacts, and as yet Carbon Capture and Storage (CCS) is untried on large scale. Moreover it is only an interim technical fix: there may be only relatively limited space for reliable storage, so it is not an option that would allow continued fossil fuel use in the long term. An easier and already widely adopted interim approach is to switch from burning coal to burning less carbon-intense gas in modern Combined Cycle Gas Turbines (CCGTs). That can almost halve emissions/kWh of energy produced, although of course that is a 'one off' saving. Once done, this transition cannot be repeated, and gas reserves are more limited than coal reserves.

A more advanced idea is to recycle some of the heat that is otherwise wasted by the production of electricity in conventional power plants. That way more useful energy can be obtained, so net emissions are less. So-called Combined Heat and Power (CHP) 'cogeneration' plants, supplying heat and as well as power, are already widely used and can raise overall energy conversion efficiency from 30% to 70% or maybe more. That could be an important part of the immediate future, feeding heat to local district heating networks, as we will see later. But there are limits: district heating makes sense only in high-density urban or possibly suburban areas.

DOI: 10.1057/9781137584434.0004

These options, although helpful, are only temporary measures. If these plants use fossil fuels there are still carbon emissions. And, like uranium, fossil fuels are finite resources. Once they have been used, they are gone forever. While avoiding energy waste and improving the efficiency of energy production and use, there is also a need to switch to non-fossil renewable fuels. Fortunately, there are plenty, from multiple sources, mostly solar derived, including direct solar energy, biomass, wind and wave energy, and hydro power, along with non-solar tidal and geothermal energy (Elliott, 2013b).

1.7 Green energy scenarios

The energy that the earth receives from the sun and other natural non-fossil sources is more than could ever conceivably be needed, although not all of it can be accessed easily, and, hydro apart, the technologies needed to turn what is available into useful forms on a significant scale have been seriously explored only in recent years. Like BP (quoted earlier), some say renewables cannot deliver enough energy to replace fossil fuel, certainly not quickly (Trainer, 2010; Smil, 2012). However, in the early 2010s, a flurry of studies emerged suggesting that, by around 2050, renewables could supply up to near 100% of all the electricity and possibly all the energy needed in the EU and in the world (EREC, 2010; ECF, 2010; PWC, 2010; WWF, 2011; Jacobson and Delucchi, 2011). Many more studies have emerged since, as we will see later, focusing on specific countries.

Some even claim that it can be done faster in some countries, given the political will (ZCB, 2014). For example, Germany is aiming to get 80% (of electricity) from renewables by 2050 and Denmark 100% (of all energy) by then. It is worth noting that around 60 countries already get over 50% of their electricity from renewable hydro, some nearly 100%. There is over 1,000 GW of hydro capacity, large and small, globally. But the so-called new renewables (wind and solar especially) are catching up, driven by falling prices. Wind is nearing 400 GW globally, solar photovoltaics (PV) 200 GW.

Given that wind and sunlight are variable, weather-dependent, resources, their total annual output is less/GW installed than from the 340 GW or so of currently active nuclear capacity. Typically, the so-called load factor for on-shore wind turbines is around 30%, some

DOI: 10.1057/9781137584434.0004

more, some less (and offshore up to 45%), that is, over a year, depending on location, they can deliver around 30–45% of the amount of electricity that they could theoretically produce if they could run continually at maximum output. The figure for solar PV is in the range 10–15%, depending on location. Nuclear plants are claimed to have load factors in the range 70–90%, or even higher, for example, for upgraded plants/new designs. But in practice they may not always achieve these high levels. The average load factor for the UK nuclear fleet for 2007–2012 was 62%. Moreover, given the rapid growth of wind power, it is beginning to challenge nuclear in energy output terms, despite wind power's relatively lower load factor. For example, in China, wind plant output has overtaken that from nuclear and has helped China to get about ten times more energy from renewables, hydro included, than from nuclear. In the United Kingdom, the output from renewables overtook that from nuclear in 2015, supplying over 19% of its annual electricity requirement (DECC, 2015).

The various '100% by 2050' renewable scenarios all involve rapid ramp up of renewables and energy efficiency, and in doing so open up many issues. Should all credible renewables be expanded equally, or should a more selective approach be adopted? If so, on what basis? Their current or projected costs and operational reliability, their potential environmental impacts and safety, or the ease of deployment at small or large scale? This book explores these strategic supply-side issues and also their relation to changes on the demand side. There are choices. There is no one 'green' energy future. Instead there are many, each with different mixes of supply and of demand management.

It is increasingly argued that there is some urgency. For example, the third US National Climate Assessment, commissioned by the White House, says that 'climate change, once considered an issue for a distant future, has moved firmly into the present', noting that extreme weather events had increased in the past 50 years, with prolonged periods of heat, floods and droughts in some areas. Temperatures may rise '2° F to 4° F more in most areas of the US in the next few decades', but reductions in some 'short-lived human-induced emissions' could cut some of the projected warming. It warned that 'the amount of warming projected beyond the next few decades is directly linked to the cumulative global emissions of heat-trapping gases and particles,' but said 'there is still time to act to limit the amount of change and the extent of damaging impacts' (NCA, 2014).

DOI: 10.1057/9781137584434.0004

1.8 Conclusion

There may not yet be a full understanding of the causes, extent and likely impacts of climate change, but to do nothing would be very risky. Fortunately, while there may be costs, should climate change turn out not to be a major issue, most of the proposed technological responses will have other benefits, including reduced air pollution and economic savings from not having to use fossil fuel, the costs of which, long term, will inevitably rise. So, assuming risky options like nuclear are avoided, there could be a 'no regrets' outcome.

The cost of inaction could be very high, much higher than the cost of the 'precautionary' approach. Stern's study suggested a 10:1 ratio in terms of the proportion of Gross National Product that would have to be spent to reduce the risk (2%), as against the cost to the economy of taking the chance, if it turns out to be as bad as many expect (20%) (Stern, 2007). A sensible insurance, with benefits in any case. However, this implies a need to decide how to act and which options to choose. The chapters that follow look at the issues, starting with environmental impacts of using renewable energy. There is no point in trying to deal with climate change by switching to new technologies if they have impacts on a similar scale.

1.9 References

Berners-Lee, M. and Clark, D. (2013) 'The Burning Question', Profile Books, London: http://www.burningquestion.info/

BP (2015) 'Energy Outlook 2035', British Petroleum, London: http://www.bp.com/en/global/corporate/about-bp/energy-economics/energy-outlook.html

Brandt, A., Millard-Ball, A., Ganser, M. and Gorelick, S. (2013) 'Peak Oil Demand: The Role of Fuel Efficiency and Alternative Fuels in a Global Oil Production Decline', Environmental Science & Technology, 47 (14), pp. 8031–8041: http://pubs.acs.org/doi/abs/10.1021/es401419t

Briggs, H. (2015) 'Global CO2 Emissions "Stalled" in 2014', BBC Report on IEA Press Story, 13 March: http://www.bbc.co.uk/news/science-environment-31872460

Brook, B. and Bradshaw, C. (2014a) 'An Open Letter to Environmentalists': http://bravenewclimate.com/2014/12/15/an-open-letter-to-environmentalists-on-nuclear-energy/

DOI: 10.1057/9781137584434.0004

Brook, B. and Bradshaw, C. (2014b) 'Key Role for Nuclear Energy in Global Biodiversity Conservation', Conservation Practice and Policy, December: http://onlinelibrary.wiley.com/doi/10.1111/cobi.12433/full

Cook, J., Nuccitelli., D, Green, S., Richardson, M., Winkler, B., Painting., R, Way, R., Jacobs, P. and Skuce, A. (2013) 'Quantifying the Consensus on Anthropogenic Global Warming in the Scientific Literature', *Environmental Research Letter,* 8 024024: http://iopscience.iop.org/1748-9326/8/2/024024/article

Dana, N. (2014) 'Climate Imbalance – Disparity in the Quality of Research by Contrarian and Mainstream Climate Scientists', *The Guardian*, 11 April: http://www.theguardian.com/environment/climate-consensus-97-per-cent/2014/apr/11/climate-change-research-quality-imbalance

DECC (2014) 'Public Want Urgent Global Action to Tackle Climate Change', Press Release, Department of Energy and Climate Change, London: https://www.gov.uk/government/news/public-want-urgent-global-action-to-tackle-climate-change

DECC (2015) 'UK Energy Statistics, 2014 & Q4 2014', Department of Energy and Climate Change, London: http://www.gov.uk/government/uploads/system/uploads/attachment_data/file/416310/PN_March_15.pdf

Dupont, C. and Oberthür, S.(2015) 'Decarbonization in the European Union', Palgrave Macmillan, Basingstoke: http://www.palgrave.com/page/detail/decarbonization-in-the-european-union-/?K=9781137406828

ECF (2010) 'Roadmap 2050', European Climate Foundation, Brussels: http://www.roadmap2050.eu

Elliott, D. (ed) (2010) 'Nuclear or Not?', Palgrave Macmillan, Basingstoke: http://www.palgrave.com/page/detail/nuclear-or-not-david-elliott/?sf1=barcode&st1=9780230507647

Elliott, D. (2013a) 'Fukushima: Impacts and Implications', Palgrave Pivot, Basingstoke: http://www.palgrave.com/page/detail/fukushima-david-elliott/?K=9781137274328

Elliott, D. (2013b) 'Renewables: A Review of Sustainable Energy Supply Options', Institute of Physics Publications: http://iopsceince.iop.org/book/978-0-750-31040-6

EREC (2010) 'Rethinking 2050', European Renewable Energy Council, Brussels: http://www.rethinking2050.eu

Evans, S. (2015) 'Nuclear Power Additions "Need to Quadruple" to Hit Climate Goals, IEA Says', Carbon Brief Blog: www.carbonbrief.

DOI: 10.1057/9781137584434.0004

org/blog/2015/01/nuclear-power-additions-need-to-quadruple-to-hit-climate-goals,-iea-says/

Fox, R. (2014) 'The Environmental Case for Nuclear Power', 15 August, Oxford University Press Blog: http://blog.oup.com/2014/08/environmental-case-nuclear-power/

Heede, R. (2014) 'Tracing Anthropogenic Carbon Dioxide and Methane Emissions to Fossil Fuel and Cement Producers, 1854–2010', *Climatic Change*, 122 (1–2), January, pp. 229–241: http://link.springer.com/article/10.1007/s10584-013-0986-y

IAEA (2014) 'International Status and Prospects for Nuclear Power 2014', International Atomic Energy Agency, Vienna, August: http://www.iaea.org/About/Policy/GC/GC58/GC58InfDocuments/English/gc58inf-6_en.pdf

IEA (2015) 'Technology Roadmap for Nuclear Energy', International Energy Agency, Paris: http://www.iea.org/publications/freepublications/publication/nuclear_roadmap.pdf

IPCC (2014) 'Fifth Assessment Report', Intergovernmental Panel on Climate Change, Geneva: http://www.ipcc.ch/

IRENA (2014) 'Renewable Power Generation Costs in 2014', International Renewable Energy Agency, Abu Dhabi: http://www.irena.org/menu/index.aspx?mnu=Subcat&PriMenuID=36&CatID=141&SubcatID=494

Jacobson, M. and Delucchi, M. (2011) 'Providing All Global Energy with Wind, Water, and Solar Power', *Energy Policy*, 39 (3), March, pp. 1154–1190: http://www.sciencedirect.com/science/article/pii/S0301421510008645

Kidd, S. (2015) 'Is Climate Change the Worst Argument for Nuclear?', *Nuclear Engineering International*, 21 January: http://www.neimagazine.com/opinion/opinionis-climate-change-the-worst-argument-for-nuclear-4493537/

Klein, N. (2014) 'This Changes Everything: Capitalism vs. the Climate', Penguin/Simon and Schuster, New York: http://www.penguin.co.uk/books/this-changes-everything/9781846145056/

de Lange, W. and Carter, R. (2014) 'Sea-Level Change: Living with Uncertainty', Global Warming Policy Foundation, London: http://www.thegwpf.org/sea-level-change-living-with-uncertainty-2/

Lomborg, B. (2014) 'Climate Change Is a Problem. But Our Attempts to Fix It Could Be Worse than Useless', *The Telegraph*, 3 November: http://www.telegraph.co.uk/news/earth/environment/

DOI: 10.1057/9781137584434.0004

climatechange/11205420/Climate-change-is-a-problem.-But-our-attempts-to-fix-it-could-be-worse-than-useless.html
Maggio, G. and Cacciola, G. (2012) 'When Will Oil, Natural Gas, and Coal Peak?', *Fuel*, 98, August, pp. 111–123: http://www.sciencedirect.com/science/article/pii/S001623611200230X - cor1
Marquina, A. (2010) 'Global Warming and Climate Change', Palgrave Macmillan, Basingstoke: http://www.palgrave.com/page/detail/global-warming-and-climate-change-antonio-marquina/?K=9780230237711
McGlade, C. and Ekins, P. (2015) 'The Geographical Distribution of Fossil Fuels Unused When Limiting Global Warming to 2° C', *Nature*, 517, January, pp. 187–190: http://www.nature.com/nature/journal/v517/n7533/full/nature14016.html
NCA (2014) 'Third US National Climate Assessment', U.S. Global Change Research Program: http://nca2014.globalchange.gov/
PWC (2010) 'A Roadmap to 2050 for Europe and North Africa', PriceWaterhouse Coopers: London: http://www.pwc.co.uk/eng/publications/100_percent_renewable_electricity.html
Royal Society/NAC (2014) 'Climate Change Evidence and Causes', Joint Report by the Royal Society UK and the US National Academy of Sciences: http://royalsociety.org/uploadedFiles/Royal_Society_Content/policy/projects/climate-evidence-causes/climate-change-evidence-causes.pdf
REN21 (2014) '2014 Global Status Report', Renewable Energy Network for the 21st Century: http://www.ren21.net/
Shell (2014) 'Annual Report', Royal Dutch Shell Company: http://reports.shell.com/annual-report/2014/servicepages/downloads/files/entire_shell_ar14.pdf
Smil, V. (2012) 'A Skeptic Looks at Alternative Energy', IEEE Spectrum, 28 June: http://spectrum.ieee.org/energy/renewables/a-skeptic-looks-at-alternative-energy
Stern, N. (2007) 'The Economics of Climate Change', Report for UK Treasury: http://www.hm-treasury.gov.uk/ independent_reviews/stern_review_economics_climate_change/sternreview_index.cfm
Tol, R. (2014) 'Quantifying the Consensus on Anthropogenic Global Warming in the Literature: Rejoinder', *Energy Policy*, 73, October, pp. 701–705: http://www.sciencedirect.com/science/article/pii/S0301421514003759 - aff0005

DOI: 10.1057/9781137584434.0004

Trainer, T. (2010) 'Can Renewables etc. Solve the Greenhouse Problem? The Negative Case', *Energy Policy*, 38 (8), August, pp. 4107–4114: http://www.sciencedirect.com/science/article/pii/S0301421510002004
WNISR (2014) 'The World Nuclear Industry Status Report 2014': http://www.worldnuclearreport.org/
WWF (2011) 'The Energy Report – 100% Renewable Energy by 2050'. World Wildlife Fund for Nature: http://www.wwf.org.uk/research_centre/research_centre_results.cfm?uNewsID=4565
ZCB (2014) 'Zero Carbon Britain', Centre for Alternative Technology Report: http://www.zerocarbonbritain.com/index.php/zcb-latest-report

DOI: 10.1057/9781137584434.0004

2

Environmental Issues: Health, Safety and Social Impacts

Abstract: *The social and environmental impacts of renewables are generally low and much less than the global impacts of conventional energy sources. However, there can be local impacts in terms, for example, of land-use conflicts, visual intrusion and wild life. The energy and carbon debts associated with building renewable energy systems are low and are falling as the technology improves, and overall the 'external' costs of renewables are generally seen as lower, and the Energy Returns on Energy Invested higher than for most other energy supply options.*

Keywords: environmental impacts; health risks; land use; noise visual intrusion

Elliott, David. *Green Energy Futures: A Big Change for the Good*. Basingstoke: Palgrave Macmillan, 2015.
DOI: 10.1057/9781137584434.0005.

DOI: 10.1057/9781137584434.0005

2.1 Environmental impacts of renewables

Even if an energy source is cheap and reliable, it should not be supported if it has significant environmental and social impacts. That must surely be the lesson learned from the previous range of energy technologies. So how do renewables stack up?

Most are relatively environmentally benign, since they are based on using mostly solar-driven energy flows, rather than on releasing energy from materials extracted from the ground. However, natural energy flows and fluxes are relatively diffuse, so that, to collect significant amounts of energy, large areas of land or sea may be required.

One way to conceptualise the situation is to look at how much energy is extracted from the natural energy flow (Clarke, 2011). The extent to which extracting energy disturbs these flows and interrupts what they were doing in the ecosystem gives us some measure of likely impacts. There is a range. The most familiar example may be hydropower. The sun's heat raises water vapour into the sky, which later falls as rain, which eventually finds its way back to the sea in river flows. This energy can be harvested by small run-of-the-river turbines or by creating reservoirs behind dams to store the energy in the head of water that is created by the hydro dam, letting it out through turbines when electricity is needed. Hydro projects can be very large and the reservoirs will often be vast, inundating large areas of land. Essentially, hydro projects try to extract a large proportion of the natural energy available, so not surprisingly the impacts can be large. Large tidal barrages, which are basically low-head hydro dams trapping high tides, will have similar large impacts. The aim is to extract a large proportion of the tidal energy, in effect blocking off estuaries.

By contrast, wind turbines take only a small amount of energy from the wind flows over land or sea, and the physical impact is therefore relatively low. Even with many wind turbines in place, the impact on overall airflow would be very small compared to say a forest. Similarly for solar energy collectors. Even with vast arrays, only small amounts of the incoming solar energy can be used, and the impact of extracting it will be small, producing small local temperature and micro-climate changes. However, since solar energy is available only during the day, and the energy conversion efficiency is lower, to get the same amount of energy out, larger areas will have to be covered than with the equivalent capacity of wind turbines on wind farms. So on that basis, wind does better. Moreover, since the energy flows in winds are an indirect and

DOI: 10.1057/9781137584434.0005

more intense form of solar energy, with air being heated differentially, in effect collecting up solar energy over time, it reduces the relative area needed by wind energy devices further. On one UK estimate, a solar PV farm would take up over eight times more area than would be needed for the bases of wind turbines in a wind farm yielding the same annual energy output (MacKay, 2014).

Winds moving over the sea create waves, which can persist for some time after the wind has died down, so waves in effect store wind, and the energy intensity can be high. Even so, large devices and areas are needed to extract the energy. It is the same for tidal energy systems. The tides are created primarily by the gravitational pull of the moon, so it is *lunar* power, not solar power, although the pull of the sun does also play a role, and the combined effect produces varying, but predictable, sea-level rises and tidal ebbs and flows, with very large energy extraction potentials, and up to 1TW of barrage, lagoon and tidal stream capacity being seen as possible globally (IRENA, 2014).

As noted earlier, the impact of tidal barrages can be very large, but the impact of tidal current turbines, essentially underwater wind turbine-like devices, will be much less, since they extract only a small part of the energy in the local tidal flow. Most of it continues to flow past them. The other main non-solar renewable source is geothermal energy, the heat deep underground, which is produced by the radioactive decay of ancient rock strata. So it is in effect natural nuclear energy, although you could say the same of solar energy, since the sun is a vast nuclear fusion reactor. However, extracting energy from sunshine, or from geothermal sources, has none of the risks associated with earth-bound attempts to use nuclear sources, although, with geothermal wells, care has to be taken to avoid venting gases from underground. Heat extraction will reduce the local heat gradient, so that, after a few decades, performance may fall off and a new site will have to be used while the heat gradient is re-established. So geothermal energy is renewable, but only in phases.

The example of biomass is harder to analyse in term of energy extraction. Although it is similar in principle to artificial solar heat collection, with biomass, the sun's energy is collected and stored over wider areas and over long time spans. Plants and other bio-matter convert solar energy via photosynthesis into a stored form in biomass. That process takes time and the energy conversion efficiency is very low, much lower than for any of the artificial energy conversion systems that have been developed. So to harvest significant amounts of energy, large areas of

DOI: 10.1057/9781137584434.0005

biomass are needed. However, while the most obvious physical impact is in terms of land use, the extraction of bio-material from the ecosystem also has an impact, something that will be explored more later on.

To summarise, in most cases, extracting energy from renewable sources has low direct impacts. There are no direct emissions, biomass combustion and geothermal venting of gases apart, though hydro reservoirs may produce methane (Brown, 2014). However, as has been indicated previously, there are varying potential land-use implications, although this has to be put in perspective. While it is sometimes claimed that nuclear plants take up much less room than wind or solar farms, that ignores the fact that nuclear plants have to be serviced by uranium mines, fuel-processing plants and waste disposal sites, which also take up space. Moreover, given the safety and security risks associated with nuclear plants and materials, these facilities are usually surrounded by large 'no-access' areas, with security fencing. As a result, one analyst even claimed that, when the full nuclear fuel cycle was included, nuclear took up *more* space than renewables (Lovins, 2009).

That view is strengthened by the fact that the land within a wind farm can be, and is, used for agricultural purposes. The only area lost is the small area of the wind turbine bases, for a small grid substation and for any access road. Moreover, if it is an offshore wind farm, then no land is lost at all, any extra grid links apart. The same goes for wave and tidal stream turbines, and if PV solar is installed on existing rooftops, it takes up no extra land. PV arrays are also being put on reservoirs now, which helps to reduce evaporation losses, an important issue in hot countries. PV arrays on land do however take up space, although that has to be put in perspective. Golf courses take up far more room, but are rarely opposed. That said, care has to be taken not to use land that is of high agricultural value, although some of the land can still be used for sheep grazing and for wild flower growth, and it has been claimed that solar farm arrays can actually enhance biodiversity, not least by protecting the area from other less appropriate uses (STA, 2014).

2.2 Human and animal impacts: safety and health risks

What about safety risks? People do fall off rooftops when installing solar arrays and there have been installation and maintenance deaths associated

DOI: 10.1057/9781137584434.0005

with wind turbines, around 150 so far globally (CWIF, 2014). However, it is hard to put these risks in the same category as the risk associated with, for example, major nuclear accidents. There are disputes over the final public death toll from the Chernobyl nuclear disaster in Ukraine, but the estimates run into tens of thousands (Fairlie and Sumner, 2006). By contrast, for wind or solar, even the most catastrophic accident (a fire or a blade throw) will be unlikely to pose much of a threat to the general public. Indeed no member of the public has so far been injured by a wind farm accident, although one small light aircraft did crash into one in the United States, killing its crew. In the case of PV, while operationally there should be few problems, the manufacture of solar cells can involve the use of toxic materials and so presents an occupation health risk. Some cells also contain toxic materials and so have to be carefully disposed of when no longer used. However, these chemical risks are similar to those with the production and use/disposal of many consumer products.

The exception to the low-risk potential of renewables is hydro. Big dams can and do fail, and large numbers of people can be killed. For example, the Banqiao/Simantan dam failure in China in 1975 claimed 30,000 lives and around the world there have been many other hydro accidents with loss of life. On this basis, large hydro is sometimes put in the same category as nuclear, with similar direct death rates per kWh of electricity generated.

However, while tragically, in the case of hydro accidents, it is relatively simple to estimate the death rate (counting drowned people), with nuclear, there are disagreements not only about the death tolls from major accident, but also about the long-term impacts of smaller leaks and even the regulated emissions, for example, given that the health effects may not show up for many years and may not be attributed to radiation exposure. Similarly with fossil fuel burning and deaths from lung disease. And it is even harder to produce estimates of health impacts from climate change and air pollution in the future, and to make sensible comparisons, especially given the relatively short period for which modern (non-hydro) renewables have been used. For example, there is as yet little experience with working with offshore wave and tidal stream technology, in what is inevitably a potentially hazardous environment.

Overall, making realistic and reliable quantitative long-term comparisons across the various technologies is hard, so the reliability of some estimates is uncertain (Wang, 2011). For example, in the study mentioned earlier by Brook and Bradshaw (2014), on-land wind is seen as having

DOI: 10.1057/9781137584434.0005

much higher fatalities due to accidents (0.15/TWh) than nuclear (0.04/TWh), and solar is even worse (0.44/TWh). However, the wind and solar figures are based on relatively limited operational experience (large capacities have only recently been installed) and for nuclear there are divergent views on the long-term impacts of radiation, including low-level radiation exposure (Fairlie, 2014). Certainly, some very different rankings have emerged from other studies. For example, the EU's long-running Externe study tried to estimate the extra social and health costs of all energy systems, renewables, fossil and nuclear, but it excluded the costs of long-term climate change impacts, since they were seen as hard to estimate accurately at that time.

On that basis it ranked coal as imposing the highest extra cost, €57/MWh (more than its generation cost), then gas, tying with biomass at €16/MWh, followed by PV solar at €6/MWh. Next came hydro, tying with nuclear at €4/MWh. And finally wind was seen as imposing the lowest extra cost, €1/MWh (Externe, 2006). A recent study by Ecofys for the European Commission included estimates for climate change costs, and put nuclear external costs higher at €18–22/MWh, more than for any renewable (Ecofys, 2014).

No technology is totally benign, and clearly some renewables do have impacts, but they are generally much lower than those associated with fossil fuels, especially when the potentially large long-term impacts of climate change are taken into account. As for nuclear, while it might avoid some of that, it is arguably a little perverse, in health terms, to promote a radiation-based technology which has significant potential for long-term damage to cellular and possibly genetic material and to the health of ecosystems.

What about lower-level social impacts? Some say that wind turbines impose unacceptable noise impacts on local residents (Windbyte, 2015). There are strict controls of permitted separation distances from habitations and of noise levels, and modern wind turbines are much quieter than earlier models. On wind farm sites it is rare to hear more than a swish sound even close up, often hard to hear over the sound of wind in trees or bushes, or the noise from any nearby road. However, noise, especially at very low levels, is a subjective issue. Some people cannot sleep in a room with fridge in it. And once you are sensitised to it, a noise, however low, can become annoying. It has been argued that people who are unhappy with wind farms for other reasons (e.g., visual intrusion) may well become hyper-sensitised to noise that would otherwise not

DOI: 10.1057/9781137584434.0005

worry most people. For example, Simon Chapman, a professor of public health at the University of Sydney, says that studies have concluded that 'pre-existing negative attitudes to wind farms are generally stronger predictors of annoyance than residential distance to the turbines or recorded levels of noise' (Chapman, 2012).

However, complaints persist, with some claiming that low-frequency infrasound is produced and can have significant health impacts. Many machines create ultrasound, often at much higher levels than would be possibly experienced at a distance from wind turbines. Several studies have been carried out, including major ones covering the United States, European Union and Australia, to see if anything really was amiss (Murray, 2014; NHMRC, 2015). So far they have not found any serious problems. Nevertheless, given that some people clearly do feel they have problems, a precautionary approach seems wise. Wind farm developers may sometimes get exasperated at what can seem like unjustified complaints, but they do seek to be good neighbours (Cummings, 2012).

Visual intrusion is a more general problem. Again it is subjective. Some people love the look of wind farms, seeing them as inspiring symbols of progress to a clean energy future. Others hate them, seeing them as gross, ugly industrial intrusions, ruining treasured views, and allegedly undermining the value of nearby properties. There is the 'shock of the new' effect. Some studies have indicated that, while initially, when at planning stage, a proportion of local people oppose wind projects, they become acclimatized to them once built. Even so, some residual opposition to wind farms often remains, and that has slowed deployment in some countries. There has also been opposition to solar farms on the grounds of visual intrusion and land use conflict issues.

Clearly, there is a need for sensitive local consultation over projects like this, taking local perceptions on board (Devine-Wright, 2011). It is understandable that people who have chosen to live in rural areas may resent what they perceive as intrusions (even if in some cases these are city people with second homes), while some who are visitors may see rural areas as leisure resources. Certainly, there can be underlying rural–urban conflicts, with cities relying on rural areas for power generation, and rural areas having to accept the physical impact, although all do benefit from the energy produced. It is notable that, if local residents are given an opportunity to share directly in the economic benefits of a local project, then opposition reduces significantly. Most of the wind projects in Denmark, and many in Germany, are locally owned, for example, by

DOI: 10.1057/9781137584434.0005

local wind co-ops, and opposition is usually minimal. An old Danish proverb is sometimes used as an explanation: 'Your own pigs don't smell' (Elliott, 2003). In some cases, however, opposition to wind or solar is less about visual intrusion and location than about the technologies' costs and efficiency and more general energy policy issues, which are looked at later in this book.

Impacts on animals are also often seen as an important issue. For example, poorly sited wind turbines located in seasonal bird flocking/migration paths have in some cases proved to be particularly problematic in the past. That can be easily dealt with by avoiding wind farm location in such areas. Most birds in normal flight avoid moving objects, and modern large wind turbine blades move relatively slowly, allowing birds plenty of time to avoid them. However, some birds clearly do not, but the numbers are usually small, as many studies have found (Kraemer, 2014a). If the problem remains, then acoustic bird scarers are available. And it's worth noting that, although all animal deaths should be avoided, cats kill very many times more birds than wind turbines (Milius, 2013).

Wind turbine infrasound noise has been suggested as one explanation for the large number of miscarriages and deaths reported for minks in a farm near a new wind project in Denmark, but minks, especially in captivity in pelt farms, are prone to illness and this case seems inconclusive, although much commented on by the opponents of wind farms (Dunchamp, 2014). More substantially, wind turbine impacts have been reported on bat populations. Their lungs are especially sensitive to the pressure drop that occurs near moving turbine blades (Baerwald et al., 2008). Evidently, they are attracted to stationary or slow-moving wind turbines since they think they are trees, and are injured if they start up fully. There are ideas for resolving that, including ultra sound bat scarers (Gosden, 2014).

All offshore systems have the potential for significant effects on marine wildlife, such as dolphins, porpoise, grey seals, and wildfowl. However, studies of the impacts of the technologies used for the extraction of energy from offshore wind, wave and tidal flows have so far suggested this is minimal. Indeed, once built, these new structures in the sea seem to provide habitats for some species: crustaceans seem to like the wind turbine foundations, while sea mammals stay clear, as do fish (Lindeboomet, 2011). Certainly, compared with the impact on fish of the high-speed turbines used in hydro plants or tidal barrages, which can cause problems (sometimes avoided by building fish ramp/ladder routes, e.g., for spawning Salmon), the relatively slowly rotating

DOI: 10.1057/9781137584434.0005

free-standing tidal rotors in tidal stream turbines represent a low hazard for fish. But the first large tidal stream project, Sea Gen in an inlet in Northern Ireland, has used a sonar system to monitor the approach of sea mammals. They seem to avoid it, but if not, the turbine can be shut down while they pass by (Phys Org, 2012).

The use of solar energy does not lead to animal impact issues, any more than do windows or other glazed areas, but in the case of concentrating solar power (CSP) there have been concerns about impact on desert wildlife and, more dramatically, about the lethal impact on birds that fly into the focused solar beams near the central power towers, as at the 392 MW Ivanpah CSP plant in California's Mojave Desert. It may be that the bright light attracts insects (as street lights do), which then attract insect-eating birds that fly to their death in the focused beam. Dish and trough CSP designs should avoid the problem, since the heat focus is smaller and more contained. But for big sun tracking-mirror arrays, focused on power towers, acoustic bird scarers may be a possible remedy and a range of other remedial ideas has emerged (Kraemer, 2014b; Kraemer, 2014c; Kraemer, 2015).

2.3 Carbon impacts

One of the reasons for turning to renewables like wind, wave, tidal and solar is that they do not directly generate carbon emissions. However, no technology is totally carbon-free. All energy conversion technologies need energy for their construction and for the production of the materials they are built from. So there is an energy debt to pay back, and since most energy still comes from fossil sources, also a carbon debt. In the case of fossil fuelled plants, there are also direct carbon emissions, and even with nuclear plants, there is an energy and carbon debt associated with the mining and processing of their fuel. Renewables do not have that problem. So it is not surprising that most studies have found that their lifetime carbon debts are generally lower than for nuclear energy systems and certainly lower than for fossil fuelled plants (Sovacool, 2008).

One way to look at the carbon debt issue is to calculate 'energy returns on energy invested' (EROEIs), that is, the amount of energy needed to build and run a plant compared with the energy it produces over its lifetime. That will give an indication of the likely carbon debt, since it

DOI: 10.1057/9781137584434.0005

will mostly be fossil energy that is used for the materials used in the construction of the plants and for producing nuclear fuel.

There are a range of estimates for and views on the 'energy in to energy out' EROEIs (Gagnon et al., 2002; Gagnon, 2008). But a study by Prof. Danny Harvey collated them and put the EROEI ratios for wind turbines at up to 80:1 for good sites, and PV solar arrays at 25:1, depending on location. They were expected to improve as the technology developed. Hydro is a special case since, once built, the plant can run for hundreds of years, so EROEIs can be very high: 100–200:1 or even more for run of the river schemes. Interestingly, the EROEIs for some energy efficiency measure are also high, perhaps up to100:1 for retrofit insulation.

By contrast, the EROEI for nuclear is put at only16:1 and likely to fall, as lower grade uranium ore has to be used, down to 5:1 and maybe less. In theory, nuclear energy, or even renewable energy, could be used to power uranium mining and fuel production, so reducing carbon emission, but there would be diminishing returns as the ore grade quality fell. EROEIs for fossil fuelled plants used to be high, when fossil fuel was abundant, nearby and easy to extract, but now have fallen, 5–6.7: 1 for coal plants and 2.2:1 for gas plants, as Harvey (2010) quotes.

There are other views, for example, some studies ignore the energy debt of fuel inputs. That, predictably, disadvantages most renewables, which have no direct fuel inputs, and put nuclear in the lead, although long-runnning hydro plants came next. Unsurprisingly, there has been a continuing debate on the EROEI methodology used (Weißbach et al., 2013; Rauge, 2013; Weißbach et al., 2014).

Moreover, for some biomass resources, the carbon debt may be higher when the full life cycle is taken into account. For example, it has been claimed that burning some types of forest-derived biomass can increase the carbon dioxide levels in the atmosphere more than an equivalent coal fired plant (RSPB et al., 2012). This is because, although, while growing, biomass absorbs carbon dioxide gas, compensating for what is produced when it is burnt, this balancing process is not exact and also takes some time. Growing, harvesting, transporting and processing biomass needs energy and there can be a long delay before the new plantation has grown enough to start absorbing carbon dioxide from the air. During this period there will thus be a net excess in the atmosphere.

Making full life-cycle estimates of biomass carbon debts is hard, since it is difficult to know how far back to go in the cycle and where to draw boundaries. So there is a range of figures. But for some forest biomass

DOI: 10.1057/9781137584434.0005

(e.g., whole tree stem wood) it does seem that the carbon absorption delay period can be significant (JRC, 2013a). Essentially what is being argued is that burning trees takes a valuable carbon sink out of service for a long period. That is not the case for forestry wastes and thinnings: they will eventually rot if not collected. So some see them as a sustainable energy source. Some also see faster growing biomass and short rotation coppicing, for example, of willow, as good energy crop options, cutting the carbon re-absorption delay time. There is of course a need to consider other aspects, for example, wider biodiversity issues and impacts on local water use, and some feel that these cannot be dealt with just by proper management and regulation.

Certainly some have fundamental doubts about the value of using any biomass for energy production. Two US academics say, bluntly, 'any use of land for the production of bioenergy feedstocks is worse for climate, water quality, soil, biodiversity, and overall ecosystem health than is the always-available option of restoring land to its ecologically best use and getting energy from other (non-biomass) sources. Put another way, getting energy from wind, water, or the sun rather than from bioenergy allows society to put land to better use than growing energy crops' (Delucchi and Jacobson, 2014).

That may seem a little sweeping. Clearly, there can be problems with big ethanol schemes and vast biofuel plantations for producing vehicle fuels, opening up major food versus fuel issues. But what about bio-wastes? For example, anaerobic biogas production from local food scrap collection and farm wastes and residues? That involves no extra land use. And, as with forest wastes, it is better than letting this material rot in the open, releasing methane, a much worse greenhouse gas than carbon dioxide. Moreover, if biomass combustion is combined with carbon capture and storage systems it could be possible to have *negative* net emissions (Azar et al., 2013). But then just planting more trees and using bio-waste and biochar production to enrich soil and enhance carbon sequestration might be a better bet (Caldecott et al., 2014).

As can be seen, it is hard to keep discussion of impacts from straying into wider strategic issues concerning possible options and routes ahead. That is the topic of a later chapter, so for now suffice it to say that land-use issues and environmental impacts are likely to remain central in the years ahead, in part since, as the next chapter suggests, the cost issues may be less significant in determining choices. Most renewables are getting cheaper, in generation cost terms, becoming comparable with, or

DOI: 10.1057/9781137584434.0005

less than, fossil fuel and nuclear generation costs, so it is conceivable that what will shape choices amongst them is their relative environmental costs.

2.4 Conclusions

Renewable energy technologies, like all technologies, do have impacts, but as the examples illustrate, they are local rather than global and hopefully most can be avoided. That cannot be said of fossil energy technologies. Their use has both local (air pollution) and global (climate) impacts. Although there may be ways to reduce or store some of their emissions, there is no way to avoid them, apart from not burning fossil fuels.

Complacency about the local impacts of renewables, for example, in the case of wind farms, on birds and bats, has to be avoided. The impacts have to be studied and reduced as much as possible by sensitive choice of location and technical adjustments (Wang et al., 2015). Similarly for the marine renewable options (Bonar et al, 2015). However the impacts have to be put in perspective. Climate change and air pollution from fossil fuel burning is likely to have a very much larger impact on wild life, and humanity, than the use of renewables (Atkin, 2014; RSPB, 2015).

By their nature, renewable energy flows are diffuse and the technology for capturing energy from them has to cover relatively large areas. However, mankind has happily accepted large areas of land being used exclusively for farming, since we need food. Conflicts with that have to be avoided, but there are many areas of marginal land not usable for cultivation, as well as deserts (for solar) and the sea (for wind, wave and tidal).

Some countries have plenty of space. A study by the US National Renewable Energy Labs (NREL) suggested that to supply the entire US with electricity from PV would need 0.6% of land area, and noted that in their base case 'solar electric footprint is equal to less than 2% of the land dedicated to cropland and grazing in the United States, and less than the current amount of land used for corn ethanol production' (Denholm and Margolis, 2008). Indeed, even in the cloudy and densely populated United Kingdom, a PV trade lobbyist group has suggested that only 1% of total UK land area would be required for enough solar arrays to meet the United Kingdom's entire electricity needs, though that ignores balancing and (night-time) backup requirements (Bennet, 2012)

DOI: 10.1057/9781137584434.0005

Then there is the risk of major accidents and wider social impacts. Meeting this head on, a study by Pushker Kharecha and James Hansen, then with NASA, says that, by avoiding emissions from fossil fuels, nuclear may have already saved 1.8 million lives, and it could save many more in future (Kharecha and Hansen, 2014). But the same would be true of renewables, with arguably fewer risks. Indeed on the basis of their estimates of carbon emissions from nuclear and wind, and their relative costs, Sovacool et al have argued that using wind would save many times more lives than using nuclear (Sovacool et al, 2013).

Views like that depend on estimates of cost, and as the next chapter shows they can be contentious. Certainly, as has been indicated earlier, the extra impact costs can be hard to calculate. Especially since the total social and environmental cost of energy production and use are not just those associated with carbon dioxide release, land use or water use. The release of radioactive materials also imposes costs. So do other emissions. A paper published in *Climatic Change* looked at 'the social cost of atmospheric release', including carbon dioxide , noxious/acid sulphur and nitrogen oxides, methane and other emissions from fossil fuel burning. It put environmental damage costs at $330–970 billion p.a. for current US electricity generation, adding around 14–34 cents/kWh to the price of electricity from coal plants, and around 4–18¢ /kWh for gas plants (Shindell, 2015).

A 2013 US study, which also covered wind and solar, used the US governments 'social cost of carbon' (SCC) measure (which also includes SO2 costs), which, in the revised version it developed, put the social cost for unabated new coal plants in the range 6.2–22.6 cents/kWh and of gas plants 2.4–10.1c/kWh. Unsurprisingly, it found that, adding the social costs to standard US Department of Energy estimates for generation costs, meant that wind (at 8 cents/kWh) and PV solar (at 13.3c/kWh), clearly beat new coal plants (at 15.5–31.9c/kWh), and wind also beat new gas plants (at 8.6–16.3c/kWh), as could PV in the higher cost gas range, all depending on the discount rates and carbon costs used. Adding CCS sequestration to coal and gas plants still left wind ahead, but gas plants (at 8.9–9.8c/kWh) then came out cheaper than PV (Johnson et al 2013). That study did not look at nuclear, but the Ecofys report for the European Commission mentioned earlier did, and as already noted, put its total extra life-cycle environmental and social cost, including around €4/MWh for accidents, at €18–22/MWh (1.9–2.3c/kWh), lower than for coal or gas, but more than for any renewable (Ecofys, 2014).

DOI: 10.1057/9781137584434.0005

Although in some eco-impact studies the resultant rankings differ slightly (Hadian and Madani, 2015), and there are issues relating the material requirements of renewable energy technology (a topic discussed in Chapter 5), the overall results generally concur with the message of this chapter. The social and environmental impacts and costs of renewables are mostly low, in most cases much lower than those associated with the use of fossil fuels, and also lower than for nuclear, with some claiming that the latter, with social and environmental costs added, is the highest cost option of all (Schneider, 2015). However, while adding in the full extra 'external' costs gives a big advantage to renewables, there are still debates about the raw costs of the various energy options. That is the subject of the next chapter.

2.5 References

Atkin, E. (2014) 'How Many Birds Are Killed by Wind, Solar, Oil, and Coal?', *Climate Progress*, 25 Aug: http://thinkprogress.org/climate/2014/08/25/3475348/bird-death-comparison-chart

Azar, C., Johansson, D. and Mattsson, N. (2013) 'Meeting Global Temperature Targets—The Role of Bioenergy with Carbon Capture and Storage', *Environmental Research Letters*, 8 034004, July: http://iopscience.iop.org/1748-9326/8/3/034004/article

Baerwald, E., Edworthy, J. and Holderet, M. (2008) 'A Large-Scale Mitigation Experiment to Reduce Bat Fatalities at Wind Energy Facilities', *Journal of Wildlife Management*, 73 (7): http://www.sciencedaily.com/releases/2009/09/090928095347.htm

Bennett, P. (2012) 'If Solar Covered One Percent of the UK It Would Meet the Country's Entire Power Demand', Solar Portal, 11 October: http://www.solarpowerportal.co.uk/news/if_solar_covered_one_percent_of_the_uk_it_would_meet_the_countrys_2356

Bonor, P., Bryden, I. and Borthwick, A. (2015) 'Social and Ecological Impacts of Marine Energy Development', *Renewable and Sustainable Energy Reviews*, 47 (July), pp. 486–495: http://www.sciencedirect.com/science/article/pii/S136403211500221X

Brook, B., and Bradshaw, C. (2014) 'Key Role for Nuclear Energy in Global Biodiversity Conservation', *Conservation Practice and Policy*, December: http://onlinelibrary.wiley.com/doi/10.1111/cobi.12433/full

DOI: 10.1057/9781137584434.0005

Brown, P. (2014) 'Tropical Dams an Underestimated Methane Source', *Climate Central*, 14 September: http://www.climatecentral.org/news/tropical-dams-methane-18019

Caldecott, B., Lomax, G. and Workman, M. (2014) 'Stranded Carbon Assets and Negative Emissions Technologies', Smith School of Enterprise and the Environment, University of Oxford, Working Paper, February: www.smithschool.ox.ac.uk/research-programmes/stranded-assets/Stranded%20Carbon%20Assets%20and%20NETs%20-%2006.02.15.pdf

Chapman, S. (2012) 'The Sickening Truth about Wind Farm Syndrome', *New Scientist*, 2885, 8 October: http://www.newscientist.com/article/mg21628850.200-the-sickening-truth-about-wind-farm-syndrome.html

Clarke, A. (2011) 'Comparing the Impacts of Renewables', *International Journal of Ambient Energy*, 15 (2) (April 1994), pp. 59–72, online in 2011: http://www.tandfonline.com/doi/abs/10.1080/01430750.1994.9675632#abstract

Clarke, A. (forthcoming) 'The Environmental Impacts of Renewable Energy', *Earthscan*.

Cummings, J. (2012) 'Wind Farms and Health: It's Not Black or White', Renewable Energy World, 17 February: http://www.renewableenergyworld.com/rea/news/article/2012/02/wind-farms-and-health-its-not-black-or-white?page=1

CWIF (2014) 'Summary of Wind Turbine Accident Data to 31 December 2014', Caithness Windfarm Information Forum, Anti-wind Farm Group Data: http://www.caithnesswindfarms.co.uk/accidents.pdf

Delucchi, M. and Jacobson, M. (2013) 'Meeting the World's Energy Needs Entirely with Wind, Water, and Solar Power', *Bulletin of the Atomic Scientists*, 69 (4), pp. 30–40: http://bos.sagepub.com/content/69/4/30.full

Denholm, P. and Margolis, R. (2008) 'Land-Use Requirements and the Per-Capita Solar Footprint for Photovoltaic Generation in the United States', *Energy Policy*, 36 (9), September, pp. 3531–3543: http://www.sciencedirect.com/science/article/pii/S0301421508002796

Devine-Wright, P. (2011) 'Renewable Energy and the Public', London: Earthscan: http://www.routledge.com/books/details/9781844078639/

Dunchamp, M. (2014) 'Turbines on Trial: Animal Miscarriages in Denmark', Master Resources, 13 June: http://www.masterresource.org/2014/06/health-effects-from-wind-turbines/

DOI: 10.1057/9781137584434.0005

Ecofys (2014) 'Subsidies and Costs of EU Energy', Ecofys Consultants Report for the European Commission: http://ec.europa.eu/energy/studies/doc/20141013_subsidies_costs_eu_energy.pdf

Elliott, D. (2003) 'Energy, Society and Environment', London: Routledge: http://www.routledge.com/books/details/9780415304863/

Externe (2006) 'External Costs of Energy', EXTERNE Reports, European Commission: http://www.externe.info/externe_2006/

Fairlie, I. (2014) 'A Hypothesis to Explain Childhood Cancers Near Nuclear Power Plants', *Journal of Environmental Radioactivity*, 133 (July), pp. 10–17: http://www.sciencedirect.com/science/article/pii/S0265931X13001811

Fairlie, I. and Sumner, D. (2006) The Other Report on Chernobyl (TORCH), Greens/EFA Group of the European Parliament: http://www.greens-efa.org/cms/topics/dokbin/118/118499.the_other_report_on_chernobyl_torch@en.pdf

Gagnon, L. (2008) 'Civilisation and Energy Payback', *Energy Policy*, 36 (9), pp. 3317–3322: http://www.sciencedirect.com/science/article/pii/S0301421508002401

Gagnon, L., Bélanger, C. and Uchiyama, Y. (2002) 'Life Cycle Assessment of Electricity Generation Options; the Status of Research in Year 2001', *Energy Policy*, 30 (14), pp. 1267–1278: http://www.sciencedirect.com/science/article/pii/S0301421502000885

Gosden, E. (2014) 'Bats Lured to Deaths at Wind Farms "Because They Think Turbines Are Trees"', *The Telegraph*, 29 September: http://www.telegraph.co.uk/news/earth/energy/windpower/11128511/Bats-lured-to-deaths-at-wind-farms-because-they-think-turbines-are-trees.html

Hadian, S. and Madani K. (2015) 'A System of Systems Approach to Energy Sustainability Assessment: Are All Renewables Really Green?', *Ecological Indicators*, 52 (May), pp. 194–206: http://www.sciencedirect.com/science/article/pii/S1470160X14005640

Harvey, D. (2010) 'Carbon Free Energy Supply', London: Earthscan: http://www.routledge.com/books/details/9781849710732/

IRENA (2014b) 'Ocean Energy', International Renewable Energy Agency, Abu Dhabi: http://www.irena.org/menu/index.aspx?mnu=Subcat&PriMenuID=36&CatID=141&SubcatID=445

Johnson, L., Yeh, S. and Hope, C. (2013) 'The Social Cost of Carbon: Implications for Modernizing Our Electricity System', *Journal of Environmental Studies Science*, 3 (4), pp. 369–375, September: http://link.springer.com/article/10.1007%2Fs13412-013-0149-5

DOI: 10.1057/9781137584434.0005

JRC (2013) 'Carbon Accounting of Forest Bioenergy', European Commission Joint Research Centre, Ispra: http://iet.jrc.ec.europa.eu/bf-ca/sites/bf-ca/files/files/documents/eur25354en_online-final.pdf

Kharecha, P and Hanson, J. (2013) 'Prevented Mortality and Greenhouse Gas Emissions from Historical and Projected Nuclear Power', *Environmental Science Technology*, 47 (9), pp. 4889–4895: http://pubs.acs.org/doi/abs/10.1021/es3051197?source=cen

Kraemer, S. (2014a) 'New Research Improves on Earlier Bird-Killing Turbine Studies', Renewable Energy World, 19 September: http://www.renewableenergyworld.com/rea/news/article/2014/09/new-research-improves-on-earlier-bird-killing-turbine-studies

Kraemer, S. (2014b) 'How to Protect Birds from CSP Towers', *CSP Today*, 6 June: http://social.csptoday.com/technology/how-protect-birds-csp-towers

Kraemer, S. (2014c) 'Preventing Bird Deaths at Solar Power Plants', Renewable Energy World, Parts 1 and 2, 11/12 September: http://www.renewableenergyworld.com/rea/news/article/2014/09/preventing-bird-deaths-at-solar-power-plants-part-1

Kraemer, S. (2015) 'Solar Flux Solution Brightens Future of Concentrated Solar Power', Renewable Energy World: http://www.renewableenergyworld.com/rea/news/article/2015/04/solar-flux-solution-brightens-future-of-concentrated-solar-power

Lindeboom, H. et al (2011) 'Short-Term Ecological Effects of an Offshore Wind Farm in the Dutch Coastal Zone: A Compilation', *Environmental Research Letters,* 6 035101: http://iopscience.iop.org/1748-9326/6/3/035101

Lovins, A. (2009) 'Four Nuclear Myths', Rocky Mountain Institute, Colorado: www.rmi.org/images/PDFs/Energy/2009-09_FourNuclearMyths.pdf

MacKay, D. (2014) 'Shale Gas in Perspective', Sustainable Energy – Without the Hot Air Blog: http://withouthotair.blogspot.co.uk/2014/08/shale-gas-in-perspective.html

Milius, S. (2013) 'Cats Kill More than One Billion Birds Each Year', *Science News*, January 29: http://www.sciencenews.org/article/cats-kill-more-one-billion-birds-each-year

Murry, J. (2014) 'MIT: Wind Farms Do Not Make You Sick', Business Green, December: www.businessgreen.com/bg/news/2385269/mit-wind-farms-do-not-make-you-sick

NHMRC (2015) Statement and Information Paper: Evidence on Wind Farms and Human Health, Australian Government National Health

DOI: 10.1057/9781137584434.0005

and Medical Research Council, Canberra: http://www.nhmrc.gov.au/guidelines-publications/eh57

Phys Org (2012) 'First Seabed Sonar to Measure Marine Energy Effect on Environment and Wildlife', PhysOrg, 10 July: http://phys.org/news/2012-07-seabed-sonar-marine-energy-effect.html#jCp

Raugei, M. (2013) 'Comments on "Energy intensities, EROIs (Energy Returned on Invested), and Energy Payback Times of Electricity Generating Power Plants" – Making Clear of Quite Some Confusion', *Energy*, 59 (15), September, pp. 781–782: http://www.sciencedirect.com/science/article/pii/S0360544213006373?np=y - affi

RSPB (20015) 'Wind Farms', Royal Society for the Protection of Birds Policy Statement, undated web site entry: https://www.rspb.org.uk/forprofessionals/policy/windfarms/

RSPB, FoE, Greenpeace (2012) 'Dirtier than Coal? Why Government Plans to Subsidise Burning Trees Are Bad News for the Planet', Friends of the Earth, Greenpeace and the Royal Society for the Protection of Birds, 2012: http://www.rspb.org.uk/Images/biomass_report_tcm9-326672.pdf

Schneider, M. (2015) 'Nuclear Power Is Risky and Unprofitable', DW Interview, 9 March: http://www.dw.de/nuclear-power-is-risky-and-unprofitable/a-18299318; see FOES/DW cost chart.

Shindell, D. (2015) 'The Social Cost of Atmospheric Release', *Climatic Change*, 25 February: http://link.springer.com/article/10.1007%2Fs10584-015-1343-0

Sovacool, B. (2008) 'Valuing the Greenhouse Gas Emissions from Nuclear Power: A Critical Survey', *Energy Policy*, 36 (8), August, pp. 2950–2963: http://www.sciencedirect.com/science/article/pii/S0301421508001997

Sovacool, B., Parenteau, P., Ramana, M., Valentine, S., Jacobson, M., Delucchi, M. and Diesendorf, M. (2013) Comment on 'Prevented Mortality and Greenhouse Gas Emissions from Historical and Projected Nuclear Power', *Environmental Science & Technology*, 47 (12), May, pp. 6715–6717: http://pubs.acs.org/doi/abs/10.1021/es401667h

STA (2014) Biodiversity Guidelines, UK Solar Trade Association: http://solar-trade.org.uk/media/140428%20STA%20BRENSC%20Biodiversity%20Gudelines%20Final.pdf

DOI: 10.1057/9781137584434.0005

Wang, B. (2011) 'Deaths per TWH by Energy Source', Net Big Future Web site: http://nextbigfuture.com/2011/03/deaths-per-twh-by-energy-source.html

Wang, B., Wang, S. and Smith, P. (2015) 'Ecological Impacts of Wind Farms on Birds: Questions, Hypotheses, and Research Needs', *Renewable and Sustainable Energy Reviews*, 44 (April), pp. 599–607: http://www.sciencedirect.com/science/article/pii/S1364032115000416

Weißbach, D., Ruprecht, G., Hukea, A., Czerskia, K., Gottlieba, S. and Husseina, A. (2013) 'Energy Intensities, EROIs (Energy Returned on Invested), and Energy Payback Times of Electricity Generating Power Plants', *Energy*, Vol. 52, 1 April, pp. 210–221: http://www.sciencedirect.com/science/article/pii/S0360544213000492#aff4

Weißbach, D., Ruprecht, G., Hukea, A., Czerskia, K., Gottlieba, S. and Husseina, A. (2014) Reply on 'Energy Intensities, EROIs (Energy Returned on Invested), and Energy Payback Times of Electricity Generating Power Plants', *Energy*, 68 (15 April), pp. 1004–1006: http://www.sciencedirect.com/science/article/pii/S0360544214001601

Windbyte (2015) 'UK Anti-wind Farm Group Web Site, Noise Section': http://www.windbyte.co.uk/noise.html

DOI: 10.1057/9781137584434.0005

3

Economic Issues: Green Energy Costs and Support Options

Abstract: *Renewable energy costs are falling, in some case dramatically, so that some are now competitive with conventional energy sources. Subsidy systems have helped them develop and will still be needed for most for a while, especially if subsidies continue to be offered to conventional energy technologies. But wind and PV solar are moving rapidly down their learning curves, and the other renewable options should follow. Guaranteed price Feed-in Tariffs have been the most successful support option, in some cases enabling consumers to buy into generating and exporting power themselves, thus changing the nature of the energy market. However, concerns about the need to maintain competitive pressures have led to new more market-orientated support schemes, for example, involving contract auctions.*

Keywords: contract auctions; costs; Feed-in Tariffs; learning curves

Elliott, David. Green Energy Futures: *A Big Change for the Good*. Basingstoke: Palgrave Macmillan, 2015.
DOI: 10.1057/9781137584434.0006.

3.1 Falling renewable costs

Renewable energy will get nowhere if it is too expensive. So far, most new renewable energy technologies have had relatively high generation costs and required subsidies to get them established. In itself, that is no reason to treat them as a failure. Most new technologies face the same challenges and some, including it seems nuclear power (despite being far from a new technology), have yet to overcome them. However, most renewables do seem to be moving to lower costs.

This is not a recent development. In its 2011 report on 'Deploying Renewables' the International Energy Agency (IEA) said that 'a portfolio of renewable energy technologies is becoming cost-competitive in an increasingly broad range of circumstances, in some cases providing investment opportunities without the need for specific economic support' (IEA, 2011).

Moreover the situation has now moved on, so that, in its report 'Renewable Power Generation Costs in 2014', the International Renewable Energy Agency (IRENA) said the cost of generating electricity from renewable sources had reached parity and even dropped below the cost of fossil fuels for many technologies in many parts of the world. It concluded that biomass, hydro, geothermal and onshore wind are all often competitive with or cheaper than coal, oil and gas-fired power stations, even without financial support and despite falling oil prices. Solar PV, with, in 2014, 180 GW in use globally, was leading the cost decline, with PV module costs falling 75% since the end of 2009 and the cost of electricity from utility-scale PV falling 50% since 2010 (IRENA, 2014b).

In many countries, including within Europe, onshore wind, with around 360 GW globally in 2014, has become one of the most competitive sources of new electricity capacity. IRENA noted that individual wind projects were consistently delivering electricity for $0.05/kWh without financial support, compared to $0.045–0.14/kWh for fossil fuel power plants. The average cost of wind energy ranged from $0.06/kWh in China and Asia to $0.09/kWh in Africa. North America also has competitive wind projects, with an average cost of $0.07/kWh. The most competitive utility-scale solar PV projects are, it claimed, delivering electricity at $0.08/kWh without financial support, and lower prices are possible with low financing costs. PV costs in China, North and South America were within the range of fossil electricity and were

DOI: 10.1057/9781137584434.0006

dropping rapidly in the Middle East, with a recent tender in Dubai, UAE, at 0.06$/kWh.

These are bold claims, but they do seem to be backed up by continuing trends. For example, in the United Kingdom, two PV solar projects were offered strike prices, under the 2014 Contracts for Difference (CfD) round, of £0.05/kWh ($0.07), near to then current UK electricity wholesale prices, three others £0.079/kWh, and a new offshore wind project planned for off Denmark has been given a contract at around £0.07/kWh ($0.10).

That is not to say that all the renewables are competitive across the board yet. Some of the figures mentioned are best prices for good sites, and gas and/or coal remain cheap competitors. In most locations, renewables are still facing well-established conventional sources, which have enjoyed, and in some cases continue to enjoy, extensive subsidies.

Table 3.1 shows a 2011 rendition of the then current and projected 2040 situation, produced for the UK government. Table 3.2 shows some 2014 US estimates for 2040 cost. They are both very conservative estimates, except arguably in the case of nuclear. Moreover, the situation

TABLE 3.1 *Generation cost estimates per MWh delivered in the United Kingdom Levelised cost in £/MWh*

Electricity option	Current cost	Cost in 2040
Onshore wind	83–90	52–55
Offshore wind	169	69–82
Tidal barrage	518	271–312
Tidal stream	293	100–140
Wave (fixed/floating)	368–600	115–300
Hydro (small/run of river)	69	52–58
Solar photovoltaics (PV)	343–378	60–90
Biomass (wastes/short rotation coppice)	100–171	100–150
Biogas (anaerobic digestion/wastes)	51–73	45–70
Geothermal	159	80
Nuclear (water-cooled reactors)	96–98	51–66
Gas with Carbon Capture and Storage	100–105	100–105
Coal with Carbon Capture and Storage	145–158	130

Note: The data in the table are not definitive: critics have suggested that the estimates for nuclear are very optimistic (being for as yet un-built plants of new types) and those for renewables pessimistic, given price reduction trends for actual projects. Certainly Mott Macdonald say they were 'bullish' on nuclear costs, and critics have argued that in fact nuclear costs are likely to go up (Harris et al., 2012). And since 2011, PV costs have fallen dramatically to be more like the 2040 estimates! Some look to new technology to reduce nuclear costs, with mini-nuclear plants being one option. In the past nuclear plant size has grown, in the hope of reducing costs: there is no certainty that going in the opposite direction will be any more successful (MacKerron and Johnstone, 2015).

Source: Mott MacDonald (2011).

DOI: 10.1057/9781137584434.0006

TABLE 3.2 *US electricity cost 2040 $/MWh*

Conventional coal	87.0
Integrated Coal-Gasification Combined Cycle (IGCC)	99.7
IGCC with CCS	121.2
Combined Cycle (gas)	81.2
Advanced Combined Cycle	77.8
Advanced CC with CCS	103.0
Combustion Turbine	148.5
Advanced Combustion Turbine	115.8
Advanced nuclear	83.0
Geothermal8	6708
Biomass	97.0
Wind	73.1
Wind offshore	170.3
Solar PV	110.8
Solar thermal	204.3
Hydroelectric	84.6

Note: Levelised cost of electricity, with grid costs, 2012$, US Department of Energy, EIA (2014) This seems optimistic on nuclear, pessimistic on offshore wind & PV.

is changing rapidly, so that many of these 'official' estimates now look dated. It is also important to note that estimates vary by country, given differing resource availability, investment and labour costs, as is shown in Wikipedia's convenient, if somewhat dated, assembly of estimates from around the world (Wikipedia, 2015).

As can be seen from the tables, onshore wind comes out as being competitive with almost all else both now and by 2040, but PV solar remains relatively expensive, as does offshore wind in the US table. That already looks dated. Onshore wind projects in the United States have been getting contracts at well under $50/MWh (£33). And as noted, in the United Kingdom, PV projects have been offered £50–79/MWh CfD strike prices (DECC, 2015). In parallel, CfD wind projects are going ahead at under £80/MWh onshore and under £120/MWh for offshore sites, with cost having fallen by 11% since 2011. In the long term, the UK Energy Technologies Institute has suggested that advanced offshore wind projects, using the best floating systems, could get the cost down to £100/MWh by 2020, £85 by 2025, £64 by 2030 and £51/MWh by 2050 (ETI, 2014). Although that may prove pessimistic, as noted earlier, a Danish offshore wind project is going ahead at £70/MWh.

In making economic comparisons, as in the tables, care has to be taken over the framework used. There is an often large difference between retail price (what consumers pay) and wholesale cost (for the generators), reflecting transmission and distribution cost, maintenance and other overheads, profits and any energy or carbon taxes imposed, these usually being passed on to consumers. More generally there are the costs of borrowing money to build plants, with interest on loans having to be paid while the plant is being built, and before it is earning any income. To exclude that, use is sometimes made of so-called overnight costs, as if the plant could be built instantly, but of course that depiction artificially advantages plants with long construction times, like nuclear plants (which can take ten years to complete), as opposed to solar, which can be installed in weeks, or wind farms, which can be installed in a few months. An alternative approach, so-called levelised costing, spreads the full construction cost over the life of the plant, as mentioned in Tables 3.1 and 3.2. That too has problems. Interest rates and costs vary over time, so making guesses and projecting them over the plant's full life can lead to errors, and in any case, most projects in reality pay off their investment costs as soon as they can. Wind and PV projects can pay off their cost in months, unlike big capital intense plants.

To some extent it is also unfair to ask renewables to be able to compete with fossil fuels, given that the latter have large environmental and health costs usually not reflected as yet in the price of energy. Some estimates were looked at earlier. IRENA says that, when damage to human health from fossil fuel use is considered in economic terms, along with the impact cost of CO_2 emissions, the price of fossil-fuel-fired power generation rises by between \$0.07 and \$0.19/kWh. There are also local development and regional social equity issues. IRENA says that for 1.3 billion people worldwide without electricity, renewables are already the cheapest source of energy and renewables also offer massive gains in cost savings and security for islands and other isolated areas reliant on expensive and polluting diesel.

IRENA also claim that renewables are competitive even when integrating high shares of variable renewables into the electricity grid. It says that 'additional spinning reserve to meet voltage fluctuations, to allow for intermittency and provide the capacity to ride out longer periods of low sunshine or wind, can add to overall system costs' but, although they depend on a range of factors, including location and market structure, 'estimates of these costs ... are in the range of \$0.035-0.05/kWh with variable renewable penetration of around 40%' (IRENA, 2014b).

Backup grid support/backup costs have certainly often been portrayed as the Achilles heal of variable renewables, but IRENA concludes, 'when

DOI: 10.1057/9781137584434.0006

the local and global environmental costs of fossil fuels are taken into account, grid integration costs look considerably less daunting, even with variable renewable sources providing 40% of the power supply. In other words, with a level playing field and all externalities considered, renewables remain fundamentally competitive.' Indeed the IEA even says that, with up to 45% penetration, backup/grid balancing might add 10–15% to costs, given technology development and higher carbon prices, in time 'the extra system costs of such high shares of variable renewable energy could be brought down to zero' (IEA, 2014a).

The specifics of grid balancing will be looked at in Chapter 4, but as can be seen, as far as IRENA and the IEA are concerned, it is not a major economic issue. Getting support for renewable expansion, however, still is. There is nothing like a level-playing field.

3.2 Support options

The starting point for any discussion of support systems for the new energy options must be that the existing energy options already enjoy massive subsidies. The Overseas Development Institute says that, globally in 2011, for every $1 spent to support renewable energy, another $6 were spent on fossil fuel subsidies (ODI, 2013). Similarly, according to the IEA, governments pumped over $0.5 trillion into subsidies for oil, gas and coal in 2012 globally, six times more than for renewables (IEA, 2012). Nuclear power has, if anything, been even more heavily subsidised, most obviously in terms of R&D spending, with, in the EU, around 78% of energy supply related R&D funding in the period 1974–2007 going to nuclear. Historic investment support for nuclear projects in the EU has also been high, at up to €8 billion per annum, compared with €5 billion per annum for coal (Ecofys, 2014). The global picture is similar. Although support for nuclear R&D has fallen off in recent years (as has most R&D), nuclear had the lion's share of energy R&D funding in IEA member countries for many decades, up to 80% in the 1970s (IEA, 2007).

While subsidies like these have continued to outpace those for renewables, some of the subsidies renewables received have been very effective at getting costs down and building up capacity. So it is tragic that there have of late been moves to reduce them. Launching its third annual Medium-Term Renewable Energy Market Report, Maria van der Hoeven, IEA director, said governments should hold their nerve:

DOI: 10.1057/9781137584434.0006

> Renewables are a necessary part of energy security. However, just when they are becoming a cost-competitive option in an increasing number of cases, policy and regulatory uncertainty is rising in some key markets. This stems from concerns about the costs of deploying renewables. Governments must distinguish more clearly between the past, present and future, as costs are falling over time. Many renewables no longer need high incentive levels. Rather, given their capital-intensive nature, renewables require a market context that assures a reasonable and predictable return for investors. (IEA, 2014b)

The big success story has been the Feed-in Tariff (FiT) guaranteed price systems introduced across the EU, and then in some other places (e.g., Japan). They have carried all before them, leading to rapid expansion of renewable capacity, wind power especially. However, in some cases, for example, with PV solar, as take-up boomed faster than expected, this approach initially led to high cost pass through to all consumers.

So in a recession and with PV costs also falling, in part due to the success of the FiTs in building an expanding market, FiT levels have been cut back across the EU. You could see this as a success story. Although consumers have certainly paid substantial amounts, FiTs had helped prices fall, so they were now less needed, although the scale of the cut cutbacks in some countries, in some cases retrospectively, led to bitter reactions from those who had invested in PV projects.

A US National Renewable Energy Labs report says FiTs were developed

> when the cost of renewable energy technologies was significantly higher than both conventional electricity prices and utilities' avoided generation costs. As renewable technology costs continue to fall and conventional fuel prices continue to rise, these policies are being adapted to these new power sector economics. In Germany, for example, the levelized cost of solar energy is now significantly below the retail price of electricity. In other regions such as in the Caribbean and the Pacific Islands, renewable electricity sources are increasingly competitive at the wholesale level as well, undercutting the avoided cost of generation from diesel or heavy fuel oil ... partly in response to these changing cost dynamics, policymakers in certain jurisdictions are beginning to introduce policies that do not fit neatly into the 'traditional' policy categories. (NREL, 2015)

The United States itself has a mixed system, a federal tax credit system nationally and Renewable Portfolio Standard (RPS) targets set in most states. A recent MIT report commented unfavourably on them both and said that it was 'not obvious why the output quota or RPS approach is so

DOI: 10.1057/9781137584434.0006

popular in the United States when experience internationally has made it so unpopular elsewhere', and that there was 'no general economic reason to favor a quantity-oriented approach like RPS over the price-oriented approaches generally used internationally' (MIT, 2015). The NREL report looked at some examples of the newly emerging schemes elsewhere, including some combining FiTs and contract auctions, and premium market approaches, as being developed in Germany and recommended by the EU as a competitive market replacement for fixed-price FiTs across the EU from 2017.

Like the US RPS, auctions in theory encourage price competition. However, the NREL report notes, they also 'tend to favour large players that are able to afford the associated administrative and transaction costs'. So some of the NREL's conclusions are what you might expect: 'the case of France demonstrates that retaining a FiT for smaller project sizes can help drive significant investment in projects typically owned by individual citizens or residents'. And, in France, it found that 'the move to auctions resulted in higher per kWh payments for generators, rather than lower prices'.

That was certainly the experience with the NFFO auction system in the United Kingdom and also its certificate trading replacement, the Renewables Obligation (RO), with UK consumers being charged more per kWh of electricity produced than consumers elsewhere under FiTs. Under the guaranteed-price FiTs, developers could borrow investment capital at low rates, whereas with the RO's certificate trading system, income varied unpredictably, and investment capital cost more, pushing up the price that had to be charged (Mitchell et al., 2006). The impacts of the United Kingdom's new Contracts for a Difference (CfD) system have yet to become clear, but it too involves competitive contract auctions for renewables. Some say that in developing countries, with fewer affluent consumers, auctions may be better than FiTs (Shrimali et al., 2015), but a combination, plus set-up grants, may be best.

Whatever the system, these subsidies do have to be paid for. That is often a politically charged issue, whether the funding comes from taxpayers or consumers, via levies or surcharges, as with FiTs or the RO. However, the costs have to be put in perspective. The UK government says that in 2013 the various renewable energy subsidies added about 3% to power bills. Fossil price rises were far larger. As renewables expand, these support cost will initially rise, but they should fall in time as the technology and market develops.

DOI: 10.1057/9781137584434.0006

When will the need for subsidies end? In a report on grid price parity, international consultants Poyry say retail grid parity will be reached sooner than wholesale parity and in some countries has already been reached, but they focus on the tougher issue of wholesale grid price parity. They claim that PV solar will get there first, ahead of wind, led, in the EU, by the south. Spain will, they claim, reach PV wholesale parity as early as 2021 followed by Portugal (2022) and Italy (2025–2032 depending on region). As for wind, they say Ireland will achieve grid parity in 2020 followed by the United Kingdom in 2021, due to high achievable onshore wind load factors in these countries, but elsewhere later. However, Turkey gets to grid parity for PV in 2018 and for onshore wind in 2019, ahead of the rest, due to higher wholesale electricity prices in the country (Poyry, 2014).

Poyry concludes: 'A system where wind and solar become competitive with wholesale market prices will mark a massive shift in the evolution of these technologies'. However, it warns that overall across the board wholesale grid-parity for renewables 'remains a long way off, and unless there is a further shift in capital or deployment costs, most large-scale renewables deployment in the next 20 years will remain subsidised'.

This seems a bit pessimistic given the examples Poyry and IRENA provide of cost reductions for specific technologies and locations, but it may nevertheless prove hard to wean developers off subsidies. The FiT system includes an automatic annual price reduction ('degression') mechanism meant to match expected market and technology trends, that is, reflecting the so-called learning curves, showing how prices fall with technical progress and market uptake. That should help the FiT level to bottom out over time.

Learning curves are useful guides to potential progress, so it is worth exploring them in a little more detail. They actually apply to any technology. As technologies develop and markets for them build, prices fall, so that there is a direct link over time between the volume of the product sold or in use and the cost of its purchase or use. The price reduction and product uptake data tend to follow exponential curves, down and up respectively, over time, and if plotted against each other on log-log graphs, a roughly straight line result, a near linear fall off. This is the so-called learning curve, still referred to as a curve despite being linear. Its slope (or 'progress ratio') will vary depending on the technology. Road gradient percentages are used.

DOI: 10.1057/9781137584434.0006

In the case of energy technologies, the plots are usually of cost per unit of energy versus total capacity installed or total energy produced. There is some debate over their reliability and a range of estimates (Jamasb and Köhler, 2008). However, the slope of the curve for PV is usually seen as the highest, typically up to 20%. It gets cheaper faster. Wind comes next at around 18%. Wave and tidal have been put at 10–15%. Most fossil-based technologies now show no sign of reduction; they have reached the end of the line, while, on some measures, nuclear has a *negative* learning curve. In part due to constant redesigns to add more safety features, it is getting more expensive, at least in the EU or United States (Boccard, 2014).

It seems that in most cases the learning curves hold true over long periods, so there is a degree of predictive ability, if enough data exists to identify the slope or other trend. However, there are limits. The curves may hold well for specific well-defined technologies, but the innovation process and market changes can throw up discontinuities and some of these may be so radical than in fact it is a new technology, with a new learning curve.

That may explain the odd results for nuclear. There have many so many completely different designs and no chance of series learning, except for the standard widely used PWR design, which some say once had a learning slope of around 5%, though now stagnating. If mass deployed, new nuclear technology might lead to breakthroughs, but, so far, there is little sign of continuity.

For PV, the result of technical change has been different, with very rapid movement down its learning curve, or curves. In theory, if technologies continue down their learning curves, they will eventually reach the point when their cost is effectively zero. In reality this might mean that they are so cheap that companies give them away, but charge for using them in some way. For example, in the consumer electronics field, mobile (cell) phones and even laptops have sometimes been given away free but with contracts required for their use. DVD players, once costing several hundred dollars, are now so cheap the market is more about the software/content and even that is now becoming available cheap online, so as the technology moves on, new markets have to be set up. How this would translate into the energy sector is unclear. PV may get so cheap, with spray on ink dye materials and 3-D printing of solar cells, it becomes a low-cost ubiquitous building material, with energy storage perhaps becoming the main cost.

DOI: 10.1057/9781137584434.0006

Certainly, the pace of change for PV solar has been very rapid, especially in Germany. In fact PV has moved down its learning curve much faster than expected, so the initial annual price degression rates adopted in Germany turned out not to be sufficient to cope, and, along with the basic FiT price offered, they had to be revised. Although the FiT cutbacks in Germany (and elsewhere) were arguably rather too abrupt and severe, and coincided with cuts imposed due to the wider economic recession (in Spain especially), overall it is clear that PV costs have fallen dramatically. Furthermore, prices look set to continue to fall.

The German think tank Agora Energiewende says that solar PV will be the cheapest form of power within a decade, at 4–6 eurocents/kWh in the EU by 2025, and 2–4 c/kWh by 2050. Or even as low as 1.5c in countries with more sun (Parkinson, 2015a). Investment bank Deutsche Bank said in 2015 that PV will be at grid price parity in up to 80% of the global market within two years, since grid-based electricity prices are rising across the world, and PV costs are still falling. Its 2015 solar outlook by leading analyst Vishal Shah predicts solar module costs will fall another 40% over the next four–five years (Parkinson, 2015b).

Moreover solar PV is not alone. Wind, the early lowest-cost winner, is still making very good progress. For some maritime countries, offshore wind will be the big resource, and as illustrated, costs are falling, but onshore wind is also getting even cheaper. A report by Lawrence Berkeley National Labs for the US Department of Energy, says that all-time low wind energy prices in the United States have increasingly enabled projects to be built in lower wind speed areas. Wind turbine prices have fallen 20–40% from their highs in 2008, with prices offered by projects to utility purchasers averaging $25/MWh for contracts in 2013 (Wiser and Bolinger, 2013).

The other renewables are also making progress. Wave and tidal stream systems are still expensive at present, but there is an expectation that some of them will follow the path pioneered by wind power and reach commercial viability within a few years. GW-level deployments are anticipated globally in the 2020s (IRENA, 2014b). Solar thermal systems are already a major success. In addition to a growing number of electricity producing concentrating solar plants (around 4 GW so far globally), and more underway (Castillo, 2014), there is over 400 GW(th) of heat supplying solar capacity in use around the world, up from 270 GW in 2012, much of it in China, offsetting expensive fossil heating fuel and

DOI: 10.1057/9781137584434.0006

dirty and increasingly scarce traditional biomass fuels (Mauthner and Weiss, 2013; REN21, 2015).

Modern biomass, for electricity and transport fuels, as well as for heat, is widely seen as a major potential growth area (bioenergy supplies around 14% of total global energy at present), although as noted in the previous chapter (and also discussed later), there are some key environmental issues with the use of some types of biomass. Similarly for large hydro. However, like some biomass options, it can be cheap. Indeed, in many cases, for well-established projects with construction costs paid off, large hydro is the cheapest source on the grid (NHA, 2015). While the prospect for more large hydro (as opposed to smaller schemes) may be limited, or at least contested, some see geothermal (12 GW of global electricity so far) as a big new option in some locations, with enhanced deep geothermal techniques making it more economic. It is already seen as competitive in parts of the United States (GEA, 2015).

Not all the renewables will prosper, though there may be surprises ahead even for some of the currently more expensive ones. What is striking about the entire renewable energy field is the extent and pace of innovation underway, with many new ideas emerging (Elliott, 2013). For most existing options, the trend is to ever lowering costs, nearing or reaching competitiveness with conventional sources, and as has been illustrated in this chapter, in some cases that has been quite dramatic. While the newly developing options will still need support on their way the lower costs, given trends like this, it could be that subsidies will no longer be required for most existing renewables, though that assumes subsidies for fossil and nuclear projects are no longer forthcoming.

That is far from clear. Fossil fuel projects continue to go ahead, some with subsidies or tax relief (e.g., for shale gas projects in the United Kingdom), as do some nuclear projects, with some still being offered more support than for renewables. For example, while new renewable energy projects in the United Kingdom are being given contracts for 15 years under the CfD system, the proposed Hinkley nuclear project has been offered a £10 billion investment loan guarantee, coupled with a 35-year contract at a guaranteed index-linked strike price of £92.5/MWh (HM Government, 2013). That is higher than wind and PV are likely to need by the time the Hinkley plant is built, if it goes ahead. Indeed, as noted earlier, under the first full round of the CfD, some wind and PV projects have already been given contracts at lower prices. By contrast, under current plans, the generous Hinkley CfD subsidy will run until

DOI: 10.1057/9781137584434.0006

2058, by which time wind and PV are likely to be very much cheaper and most other renewables should also have reached grid price parity and will need no support.

3.3 Conclusions

Divestment campaigns around the world may make it harder for fossil fuel projects to obtain funding in future. The World Bank says it will invest heavily in renewables and clean energy and only fund coal projects in 'circumstances of extreme need', where no clean option was viable at a reasonable price, because climate change will undermine efforts to eliminate extreme poverty (Kim, 2014). With some governments also indicating that they will no longer support investment in large fossil energy projects overseas, the social and environmental costs have begun to come home to roost. The World Bank does not provide funding for nuclear projects, and getting finance for new nuclear projects may also be increasingly difficult, as costs and risks rise, unless government are willing to step in, and it seems few are.

By contrast, as this chapter has shown, renewables are getting cheaper. There may be occasional divergencies from this trend, as happened in the case of wind energy in the United States when there was political uncertainty about the renewal of the federal production tax allowance. The rise in energy prices, driven by oil price increases (it went over $100/barrael in 2008/2009) also had a knock-on effect on the manufacturing cost of renewable energy technologies, especially those with high energy/materials requirements. However, these hiccups apart, the cost trend has been down and that has helped support dramatic take-up and continual market growth. For example, for wind, the global annual market grew by 44% in 2014 (GWEC).

Although in an increasingly risk adverse investment climate getting finance for any projects is hard, renewables are one of the few areas of rapid growth, with the recession producing only temporary overall reductions. The slowdowns in the West have been partly balanced by expansion in the East. Consultants Ernst and Young's 2014 retrospective study of 'Europe's Low Carbon Industries' confirmed the view that China was leading. It noted that 'concerns about future policy support in the EU and the US have delayed investment decisions since 2011. Renewable energy investments in the EU and the United States respectively

DOI: 10.1057/9781137584434.0006

decreased by 58% and 33% between 2011 and 2013. On the other hand, China's investments, which benefited from a more stable framework, increased by 8% between 2011 and 2013' (Ernst and Young, 2014).

With the recession lifting, investment in renewables globally in 2014 has been put at $270 billion, a 17% rise from the 2013 figure of $232 billion (GTR, 2015). Investment capital has come from banks and funds, less often from governments, but increasingly directly or indirectly from consumers, via FiT systems, which reward 'self-generating' private domestic investors, while passing the costs on to all users, with PV solar being the key option.

The growth of the so-called prosumer movement in Germany has been spectacular. By 2013, of the near 80 GW of renewable capacity that was in place, 35% was owned by private domestic consumers, and a further 11% by farmers. The idea is spreading. The International Energy Agency noted that in some countries 'it is now more cost-effective for households to produce their own power from PV than to purchase electricity from the grid' (IEA, 2014c). In addition, there is a growth of local energy co-operatives. For example, in Germany, there are over 940 energy co-ops, with, in 2014, 76,500 members, most of them involved with renewables, in towns and in rural areas, and over 100 rural communities have become 100% renewable energy based (DGRV, 2012; Debor, 2014).

This spread of local ownership has challenged the market power of the big energy utilities and may represent a new focus for decentralised energy development and energy system financing, with utilities, if they are to survive, moving into a servicing role (Schleicher-Tappeser, 2012). As we will be seeing later, this is already happening in Germany. That is not to say that more conventional renewable energy projects will not also go ahead, with some of them being needed to balance the overall system, but the balance between large and small, local and centralised, is changing. The next chapter looks at how that balance may play out in practice, in terms of system integration. Can a coherent system be created that balances variable renewables at various scales?

3.4 References

Boccard, N. (2014) 'The Cost of Nuclear Electricity: France after Fukushima', *Energy Policy*, 66, March, pp. 450–461: http://www.sciencedirect.com/science/article/pii/S0301421513011440

Castillo, A. (2014) 'Concentrated Solar 2014 Review', *CSPToday*, 24 December: http://social.csptoday.com/markets/2014-year-review

Debor, S. (2014) 'The Socio-economic Power of Renewable Energy Production Cooperatives in Germany: Results of an Empirical Assessment', Wuppertal Institute: http://epub.wupperinst.org/frontdoor/index/index/docId/5364

DECC (2015) 'Contracts for Difference (CFD) Allocation Round One Outcome', UK Department of Energy and Climate Change, London, February: http://www.gov.uk/government/statistics/contracts-for-difference-cfd-allocation-round-one-outcome

DGRV (2012) 'Energy Co-operatives', Deutscher Genossenschafts und Raiffeisenverband e.V., Berlin: http://xa.yimg.com/kq/groups/20593576/937211628/name/Energy_Cooperatives%20DGRV%20 2012%20how%20to%20Establish1%2Epdf

Ecofys (2014) 'Subsidies and Costs of EU Energy', Ecofys consultants' Report for the European Commission: http://ec.europa.eu/energy/studies/doc/20141013_subsidies_costs_eu_energy.pdf

EIA (2014) 'Levelized Cost and Levelized Avoided Cost of New Generation Resources in the Annual Energy Outlook 2014', Energy Information Administration, US Department of Energy, Washington DC: http://www.eia.gov/forecasts/aeo/electricity_generation.cfm

Elliott, D. (2013) 'Renewables: A Review of Sustainable Energy Supply Options', Institute of Physics Publications: http://iopsceince.iop.org/book/978-0-750-31040-6

Ernst and Young (2014) 'Europe's Low Carbon Industries: A Health Check', Ernst and Young Consultants: http://www.ey.com/FR/fr/Newsroom/News-releases/communique-de-presse---ey---Low-Carbon-Industries

ETI (2014) 'Offshore Wind Floating Platform Demonstration Project FEED Study: PelaStar Cost of Energy', UK Energy Technologies Institute, Birmingham and Loughborough: http://www.eti.co.uk/wp-content/uploads/2014/03/PelaStar-LCOE-Paper-21-Jan-2014.pdf

GEA (2015) US Geothermal Energy Association web site: http://geo-energy.org/geo_basics_plant_cost.aspx

GTR (2015) 'Global Trends in Renewable Energy Investment', UNEP/Bloomberg New Energy Finance/Frankfurt School: http://fs-unep-centre.org/publications/global-trends-renewable-energy-investment-2015

GWEC (2015) Global Wind Energy Council Data: http://www.gwec.net/global-figures/graphs/

DOI: 10.1057/9781137584434.0006

HM Government (2013) 'Initial Agreement Reached on New Nuclear Power Station at Hinkley', UK Government Press Release: https://www.gov.uk/government/news/initial-agreement-reached-on-new-nuclear-power-station-at-hinkley
IEA (2007) 'Reviewing R&D Policies', International Energy Agency, Paris: http://www.iea.org/publications/freepublications/publication/ReviewingR&D.pdf
IEA (2011) 'Deploying Renewables: Best and Future Policy Practice', International Energy Agency, Paris: http://www.iea.org/publications/freepublications/publication/deploying-renewables-2011.html
IEA (2012) 'World Energy Outlook 2012 Executive Summary', International Energy Agency, Paris: http://www.iea.org/publications/freepublications/publication/english.pdf
IEA (2014a) Speech by IEA Executive Director, at the launch of its report 'The Power of Transformation – Wind, Sun and the Economics of Flexible Power Systems', International Energy Agency, Paris: https://www.iea.org/newsroomandevents/speeches/140221_GIVAR_speech.pdf
IEA (2014b) 'Third Annual Medium-Term Renewable Energy Market Report', International Energy Agency, Paris: http://www.iea.org/w/bookshop/480-Medium-Term_Renewable_Energy_Market_Report_2014
IEA (2014c) 'RE-Prosumers', International Energy Agency, Paris, IEA-RETD Report: http://iea-retd.org/archives/publications/e-prosumers-report
IRENA (2014a) 'Ocean Energy', International Renewable Energy Agency, Abu Dhabi: http://www.irena.org/menu/index.aspx?mnu=Subcat&PriMenuID=36&CatID=141&SubcatID=445
IRENA (2014b) 'Renewable Power Generation Costs in 2014', International Renewable Energy Agency, Abu Dhabi: http://www.irena.org/menu/index.aspx?mnu=Subcat&PriMenuID=36&CatID=141&SubcatID=494
Jamasb, T. and Köhler, J. (2008) 'Learning Curves for Energy Technology: A Critical Assessment', in Grubb, M., Jamasb, T., and Pollitt, M. (eds) 'Delivering a Low Carbon Electricity System: Technologies, Economics and Policy', Cambridge University Press: http://www.cambridge.org/asia/catalogue/catalogue.asp?isbn=9780521888844&ss=exc

DOI: 10.1057/9781137584434.0006

Kim, J. (2014) World Bank President Interviewed by the Guardian, 23 November: http://www.theguardian.com/environment/2014/nov/23/world-bank-to-focus-future-investment-on-clean-energy

MacKerron, G. and Johnstone, P. (2015) 'Small Modular Reactors – The Future of Nuclear Power?', Sussex Energy Group, SPRU, University of Sussex: http://blogs.sussex.ac.uk/sussexenergygroup/2015/03/02/small-modular-reactors-the-future-of-nuclear-power/

Mauthner, F. and Weiss, W. (2013) 'Solar Heat Worldwide: Markets and Contribution to the Energy Supply', IEA Solar Heating and Cooling Programme: http://www.slideshare.net/UweTrenkner/worldwide2011-ed2013-lo-res

MIT (2015) 'The Future of Solar Energy', Massachusetts Institute of Technology, Cambridge, MA: http://mitei.mit.edu/futureofsolar

Mitchell, C., Bauknecht, D. and Connor P. (2006) 'Effectiveness through Risk Reduction: A Comparison of the Renewable Obligation in England and Wales and the Feed-in System in Germany', *Energy Policy*, 34, pp. 297–305: http://www.sciencedirect.com/science/article/pii/S0301421504002411

Mott MacDonald (2011) 'Costs of Low-Carbon Technologies', Report for the Committee on Climate Change, May: http://hmccc.s3.amazonaws.com/Renewables%20Review/MML%20final%20report%20for%20CCC%209%20may%202011.pdf

NHA (2015) US National Hydropower Association web site: http://www.hydro.org/why-hydro/affordable/

NREL (2015) 'The Next Generation of Renewable Electricity Policy', Report for the US National Renewable Energy Labs, Golden, February: http://www.nrel.gov/docs/fy15osti/63149.pdf

ODI (2013) 'Time to Change the Game', Overseas Development Institute, London: http://www.odi.org.uk/subsidies-change-the-game

Parkinson, G. (2015a) 'Solar at 2c/kWh – The Cheapest Source of Electricity', Renew Economy, 25 February: http://reneweconomy.com.au/2015/solar-2ckwh-cheapest-source-electricity-47282

Parkinson, G. (2015b) 'Solar at Grid Parity in Most of World by 2017', Renew Economy, 12 January: http://reneweconomy.com.au/2015/solar-grid-parity-world-2017

Poyry (2014) 'Is the End in Sight for Renewable Subsidies?' Poyry Consultants Overview, October: http://www.poyry.com/news/articles/end-sight-renewable-subsidies

DOI: 10.1057/9781137584434.0006

REN21 (2015) '2015 Global Status Report', Renewable Energy Network for the 21st Century, http://www.ren21.net/

Schleicher-Tappeser. R. (2012) 'How Renewables Will Change Electricity Markets in the Next Five Years', *Energy Policy*, 48, September, pp. 64–75: http://www.sciencedirect.com/science/article/pii/S0301421512003473

Shrimali, G., Konda, C., Farooquee, A. and Nelson, D. (2015) 'Reaching India's Renewable Energy Targets: Effective Project Allocation Mechanisms', Climate Policy Initiative, with the Indian School of Business, http://climatepolicyinitiative.org/publication/reaching-indias-renewable-energy-targets-effective-project-allocation-mechanisms/

Wikipedia (2015) 'Cost of Electricity by Source': https://en.wikipedia.org/wiki/Cost_of_electricity_by_source

Wiser R. and Bolinger, M. (2013) '2013 Wind Technologies Market Report,' Lawrence Berkeley National Labs Report for the US Department of Energy: http://emp.lbl.gov/publications/2013-wind-technologies-market-report

DOI: 10.1057/9781137584434.0006

4

Integration Issues: Dealing with Intermittency

Abstract: *Some renewable energy sources are variable, so that mechanisms have to be available to balance their impact on the power grid system. Grid systems already deal with supply and demand variations, in part by ramping backup plants up and down, and with moderate levels of renewables added, this will remain sufficient, but as and when the renewable contribution rises, the existing mechanisms will have to be extended and additional mechanisms added. Options include smart grid demand-management systems, supergrid imports balanced over time by exports of excess supply, increased use of pumped hydro and new compressed air storage systems, along with power-to-gas systems converting surplus supply to stored hydrogen/methane and flexible CHP/district heating systems linked to large heat stores.*

Keywords: backup plants; energy storage; power to gas; smart grids; supergrids; variable renewables

Elliott, David. *Green Energy Futures: A Big Change for the Good*. Basingstoke: Palgrave Macmillan, 2015.
DOI: 10.1057/9781137584434.0007.

4.1 Power grids and balancing renewables

The newly emerging energy system based on renewable sources will be very different from the systems that exist now around the world. In those, the emphasis has been on a few large plants sending electricity long distances down grid transmission lines to large numbers of consumers. That made sense in that large coal-fired plants could be located near coalfields or ports, and large coal plants were more efficient than small ones. When nuclear plants were added to networks in some countries, they too were large and were mostly located well away from cities for safety's sake.

The system logic began to change when small very efficient gas turbines became available. The gas grid could supply them wherever they were, often nearer to users. The advent of renewables takes that new logic to the next stage. In many cases, generation could be done near users, or even by users. That would avoid the energy losses from long-distance transmission. Perhaps 10% is lost over each 1,000 km. However, not all renewables can or should be locally sited near users. Large wind farms are best sited where the wind is strongest and most reliable, which will usually be in relatively remote upland areas, or of course offshore. The location of wave, tidal and hydro projects are inevitable geographically defined, as are geothermal projects. So there is still a need for grid transmission to link them to users, although it will be more of a matrix type network than a big central grid.

The grid can also play a key role in resolving what some see as the major problem with renewables, the fact that some of the sources are variable. For example, the winds do not always blow everywhere all the time. The grid can help balance variations in local wind availability and demand. Some areas may actually have more that they need at some point in time and can export it to areas which need energy but, at that time, have low wind. Similarly for solar. Local energy availability and local energy demands can thus be balanced, to some extent, though to get fuller balancing, longer transmission distances may be needed, across whole countries, regions and even continents. With 'supergrids', electricity is shifted, for example, across the EU, using high-voltage direct current (HVDC) links. There will be energy losses with long-distance transmission, as with conventional grids, but with HVDC they are much lower, perhaps 2% per 1,000 km. Many studies are underway of the viability of such an approach (Elliott, 2012).

DOI: 10.1057/9781137584434.0007

As can be seen, this is a long way from the idea of relying on local decentralised energy, although these options are not antagonistic: they can support each other, with local surplus feeding into the supergrid. It makes sense to share excesses around, especially from areas where the resource is large and reliable, as, in the extreme, with solar in desert locations, but this supergrid idea is only part of the solution to local renewable variability. To understand the others, it is helpful to look at how the grid system works at present. When there is a local shortage, or when a power plant fails, backup capacity is used to fill the gap. This rarely involves starting up plants from cold (that takes time), but instead uses already running plants, in what is called 'spinning reserve' mode, ramped up to meet sudden peaks in demand. That is done all the time. Every evening there are peaks and the system copes happily with them, mostly using the linked-in gas turbines. The demand surges are usually short, an hour or so at most. With variable renewables on the grid these plants would just ramp up and down more often to meet any shortfalls. There is no need for new extra backup plants, since they already exist, as do other grid balancing systems, and can be used to balance renewables, at least assuming moderate input levels (Boyle, 2009).

4.2 Grid backup and balancing

Denmark already obtains well over 30% of its annual average electricity from wind power and it seems that up to around that level, existing systems can cope quite happily, with few major changes needed. The existing backup plants are of course at present fossil fueled, mostly gas turbines. Running these plants occasionally more often to balance renewables would reduce the fuel saved and carbon emissions avoided by having renewables on the grid by a small amount. Gradually though some of these plants, or their improved replacements as they get old, could use renewable fuel to avoid the carbon emission entirely, for example, biogas. In addition, use could be made of surplus electricity from wind and solar plants to make hydrogen gas, by the electrolysis of water. This could be stored and then used to fuel the standby plants when needed. In some versions of this so-called power to gas idea now being tested, the hydrogen is converted to methane gas, using carbon dioxide captured from air or from the exhausts of fossil power plants (DENA, 2014a). Power-to-gas systems could be an increasingly logical

DOI: 10.1057/9781137584434.0007

option when moving beyond around a 30% contribution from variable renewables, since then there would be more surplus energy to convert to help with balancing. Using it in this way makes more sense than simply dumping it, although, depending on the market, it may be more attractive to export it, for example, via the supergird.

The system described in outline should be able to handle short- to medium-term variations in supply and demand, and can be augmented by importing energy, if needed and available, on the supergrid. That can reduce the need for backup, perhaps, on average, by up to half for a cross-EU supergrid system (Aboumahboub et al., 2010). There is also another form of backup, which is already used, pumped hydro. When there is surplus electricity it is used to pump water up behind a hydro dam into its reservoir. When electricity is needed, this extra head of water is released through the hydro project's turbines. Pumped hydro, using cheap off-peak electricity and supplying electricity to the grid when there is peak demand, is well established, but now it has an extra role, balancing variable renewables, for example, excess wind-derived electricity from Denmark is being exported to hydro projects in Norway and electricity imported back when there is no wind/high demand in Denmark (JRC, 2013).

As more renewables are added to the grid, extra large storage systems like this may be needed, and they, along with the backup plants and reserve capacity, could also play a role in meeting longer-term lulls in renewable availability. There are several other bulk storage options, including compressed air stored in large underground caverns, and the production and storage of liquid air, for use when needed to drive turbines. However, storage is expensive, and for the moment simple gas turbine backup plants are a cheaper way to balance both short- and longer-term variations in renewable availability. Basically storing gas, whether fossil gas or green gas, is easier and cheaper than storing electricity.

As we shall see later, there is debate as to when and if large-scale storage will be needed. There is, however, one storage option that is arguably viable now, namely, heat. Heat storage, if done with large units, is very efficient, with low losses, since the surface to volume ratio of the store goes down with size. It offers a grid balancing option. For example, gas or biomass-fired Combined Heat and Power (CHP) plants generate heat and power and feed the heat to district heating networks, or it can be stored for later use in large heat stores. The ratio of heat to electric power

DOI: 10.1057/9781137584434.0007

produced by a CHP plant can be changed quite rapidly. So if demand for power is low and/or there is too much coming from renewables, more heat can be produced, and if it is not needed at that point it can be stored, in large water tanks. Alternatively, if demand for electric power is high, CHP power output can be increased and heat output reduced, with any extra heat demand being met from the store (JRC, 2012). CHP is increasingly being biomass fired, and surplus electricity from wind/PV can also be converted into heat using immersion heaters in the stores. So emissions from the CHP system, already relatively low, are cut further.

Finally, there is yet another approach which may be the cheapest grid balancing option of them all, that is, managing demand. Some loads can be deferred without much impact on users, for example, commercial and domestic freezer units can coast without power for some hours without noticing the loss. Interactive load management systems can switch off loads automatically, subject to users' prior agreement, during peak demand times, in effect shifting the peak until later. A similar result can be obtained by the introduction of variable 'time of use' charges, with high prices being charged at peak demand times. Softer variants offer consumers online information on demand and costs, via smart meters, to allow them to plan their energy use better. The hope is that the development of 'smart grid' systems like this could change the relationship between supply and demand fundamentally, making it possible to cope with reduced input from variable renewables, and, by reducing waste, limit the extra cost (IEA, 2015).

4.3 Grid balancing: the longer term

It is sometimes claimed that plants using variable renewable sources will need full, 100%, backup, since at times there will be no energy available from them. As can be seen, in fact, at present, this backup already exists, in the form of standby plants and other grid balancing measures, including, if needed, the use of fast-acting short-term reserve plants. That is because what is sometimes called 'intermittency' is not limited to renewables. All plants can and do shut down unexpectedly, and, as described earlier, the power grid system is designed to cope with that and the daily demand variations, mainly by ramping up output from 'spinning reserve' plants, as well as, for very small variations, by varying mains frequency. As argued, this system should be able to cope with

DOI: 10.1057/9781137584434.0007

renewables up to quite high levels, aided by CHP/heat stores, smart grid demand management and, if needed, supergrid imports. Pumped hydro and other storage options could also help, especially with the possibly larger and longer loss of power issues, for example, during long wind lulls, when there may also be a need to crank up extra reserve plant to meet gaps, calling on the built-in extra plant margin that grids have for this purpose. In addition, some renewables, like geothermal and biomass, and indeed hydro, are not variable and would be available when there were long lulls in wind and solar inputs. Tidal schemes should also be able to supply inputs when wind and solar inputs are low, since the tides are unrelated to wind or solar energy, and pumped storage is also an option for tidal barrages/lagoons.

Would this type of balancing system work with very high levels of renewables? A Poyry report for the UK Committee on Climate Change spelt out how it would be possible to balance 'stretching but feasible' 2050 scenarios with high levels of renewables, supplying up to 94% of the United Kingdom's annual electricity needs (Poyry, 2011). They found that 'the electricity system was able to accommodate these high levels of renewable generation whilst complying with the specified constraints on emissions and security of supply'. However, they said 'this was at the cost of shedding low variable cost generation [i.e., some dumping of surplus power] and construction of new peaking capacity' after 2030.

By contrast, a study by the UK Pugwash group, to which I contributed, found that up to 80% of UK electricity, and possibly more, could come from renewables by 2050, without the need for any new fossil backup plants, if supergrid links and power-to-gas systems were developed for grid balancing, exporting and/or using the surplus output rather than shedding it. Using the 2050 Pathways modeling system produced by the UK Department of Energy and Climate Change, the study also found the system would be robust against variations in renewable availability over time (Pugwash, 2013). Poyry's 94% renewables scenario was similarly robust. They reported that 'in our (very) high renewable scenarios, we found that there is relatively little difference between the level of security of supply ... in an average weather year and from the level in one of our extreme weather years'.

In these high renewables scenarios, there would be no need for nuclear. Indeed it has been argued that it would just get in the way, since nuclear plants are inflexible and would not be able to ramp up and down regularly and quickly to balance variable renewables. Given that it is

DOI: 10.1057/9781137584434.0007

often argued that 'always available' base-load is vital and that nuclear can provide it, it may be worth exploring this issue in a little more depth.

While the economic case for nuclear may be weak, and in the longer term there may be fuel scarcity/carbon emission issues, some still say it should and could play a role in the interim in backing up renewables. The reality is that this is a limited option. Nuclear plants cannot vary their output rapidly or regularly without safety problems. In addition to heat stress issues with regular ramping up and down, it takes time for the activated Xenon gas that is produced, when reaction levels are changed, to dissipate. That can interfere with proper/safe reactor performance. Nuclear plants are in any case usually run 24/7, as base-load load plants, so as to recoup their large capital cost. While some can respond to daily energy demand cycles (demand peaks in the evening, low demand at night) they cannot cope with the fast irregular variations likely with wind and so on on the grid. It is conceivable that they could be used to cover the occasional longer periods when wind and so on is at minimum, but that would mean running the nuclear plants at lower level at other times, ready to ramp up slowly to meet the lull periods, which would undermine their economics.

Moreover, if there is a large nuclear contribution, and also a large renewables contribution, there can be head-to-head operational conflicts when energy demand is low, for example, at night in summer, when, in the United Kingdom, demand is around 20 GW. The United Kingdom is aiming for 16 GW of nuclear in the 2020s and more later, taking it well beyond 20 GW, and maybe 30 GW of renewables by around 2020 and more later. So, assuming all the occasional excess output cannot be exported, or stored, which plants would be turned off when demand is low? The nuclear operators do not want nuclear output to be 'curtailed'. Nor do the renewable plants: they would lose money. It would be a waste either way. The two technologies are incompatible at large scale on the same grid.

It is conceivable that nuclear plants could be run in power-to-gas mode, generating hydrogen gas when their electricity output is not needed, in which case that might be used to balance renewables. Some old plants may end up being used in this way for example, in France, which will have many more than it needs as renewables expand. However, it is unlikely that anyone would build new nuclear plants for this purpose, given their high cost and poor economics in this operating mode.

DOI: 10.1057/9781137584434.0007

It is possible that new smaller, cheaper, nuclear plants may emerge, which could be run in CHP mode, supplying heat to local users, but would everyone be happy to have mini-nuclear plants in or near cities? Even more speculatively, nuclear fusion may be viable in the longer term, post-2050, and, rather than being used for base-load, used for hydrogen production, in which case it might offer a way to balance variable renewables. However, fusion is still some way off. Certainly, even if all goes well with the current research work, it will not be available in time to deal with the urgent problem of climate change, or to help renewables to do that in the near term.

There may thus be limits to the role that nuclear might play in the new energy system, and as has been indicated, although there would still be a need for some backup plants, along with other balancing measures, the provision of 'base-load' may actually not be the central concern. As a cross EU-report for Agora Energiewende put it, looking to 2030, 'base load capacities will decrease relative to those of today, while peak load and mid-merit capacities will increase', all of which means a very different approach to system design and management: 'renewables, conventional generation, grids, the demand side and storage technologies must all become more responsive to provide flexibility' (Agora, 2015).

How far can this go? Can fossil fuel backup be avoided? Some tests have already been carried out on flexible system management in Germany, with a 'virtual' power plant system combining wind and PV solar, along with hydro pumped storage and biogas top-up plants. Using advanced system management software, it was found possible to meet varying electricity demand over a full year with little need for grid imports and no fossil backup (Barnham, 2014). Moving to a full-scale national system may require more inputs and balancing, especially if heating and transport needs are also to be met. However, as already indicated, in a comprehensive total energy system, the various scales and types of renewable would be located in and balanced by a flexible smart grid based and supergrid-linked supply and demand system, with CHP/heat stores playing a role, partly backed up by 'firm' renewables (hydro, geothermal and biomass), able to supply output on demand, and by 'power to gas' conversion of surplus from variable renewables. That was suggested for the Pugwash UK high renewables scenario mentioned earlier. Interestingly, the Pugwash 2050 non-nuclear, 80% renewables pathway (with around 70% coming from indigenous UK renewables) was not only robust in a five-day 'stress test' for viability when renewable

DOI: 10.1057/9781137584434.0007

inputs were low and demand high, it also avoided the need for new fossil backup, though it used the residual 20%. Interestingly it cost slightly less than an equivalent high nuclear scenario that was explored in parallel. Moreover, its large wind element produced significant excess electricity at times, and, it claimed, exporting this via a supergrid could earn the United Kingdom £15 billion p.a. In practice though it might not always be possible to sell it all at reasonable prices (that would depend on demand in the EU): it was envisaged that, on average, perhaps half would be used to make storable gas for grid balancing and, possibly, for heating and transport use. With extensions (including more renewables and more power-to-gas conversion), it was suggested that it should be possible get up to near 100% renewables, supplying most energy needs, although the cost might rise, depending on the success with developing new technology, with the transport sector probably being the most difficult to deal with, given that the Pugwash scenario avoided importing biofuels/biomass (Pugwash, 2013). A more recent high renewables UK study by researchers at Imperial College London, which excluded the use of biomass entirely, and focused just on electricity supply, concluded that 'using current technologies, renewable shares up to about 80% are possible without significant cost increases', although they added 'to go beyond that, improved technologies (either for dispatchable supply or for storage) and/or significantly increased interconnection and imports from beyond the UK are necessary'. Even so 'only for the scenarios with renewables above 70% of installed capacity do costs start rising above 0.10 GBP/kWh, and only for 90% and 100% renewables do the costs go significantly beyond that figure' (Pfenninger and Keirstead, 2015).

Looking many decades ahead, as in scenarios like this, is hard, given they are inevitably technologically speculative and also based on assumptions about future demand. It looks possible at some point, but it is hard to be sure when or even if non-fossil options can meet all needs in all sectors, including backup and balancing needs. So, although scenario like those mentioned do try, it is even harder to come up with cost figures for balancing that are reliable, and it may well be cost that is the defining issue. However, we might be able to extrapolate from current experience. At present these costs are shared by all generators (or rather the suppliers, who pass them on to consumers), since all plants need backup. With some old plants going off line and new variable renewable plants coming on line, the UK government has set up a Capacity Market to make sure that there will be enough capacity available for backup

DOI: 10.1057/9781137584434.0007

when and if needed. Most of the plants already exist, but the new market provides extra payment to make sure they are available for extra service if needed. Around 50 GW has been contracted in this role, including some storage plants. The cost is being passed on to the consumer and the government has said the first 50 GW of contracts would add around £11 extra to typical annual bills. So that gives an idea of backup costs now, with renewables supplying around 19% of UK electricity in 2013.

As renewables expand, grid balancing systems must also expand, and hopefully more of this will make use of non-fossil-fuelled plants, but the IEA says that even at a 45% renewable penetration, backup/balancing would still only add 10–15% to electricity costs, and it claims that this could be reduced as technology improved, and would be offset by savings from reduced fossil fuel use and associated carbon costs, so that, in the long term, the net extra cost of using variable renewables may be low or even zero. Indeed the IEA suggests that, depending on carbon cost and technology/system improvement, eventually, the net cost may even be less than for conventional systems (IEA, 2014). That may seem technologically optimistic, but, as we have seen, there is a range of types of balancing options available, some of which, on the demand side, may reduce net costs (by avoiding wasteful over-production), or in the case of supergrids, earn an extra income from exports, while overall reducing the need for backup plants and other balancing measures (Rodriguez et al., 2014).

There is much work being done in this field currently, seeking to identify optimal approaches to balancing with large renewables contributions, and cutting across many engineering disciplines (Apt and Jaramillo, 2014; Jones, 2014; Sorensen, 2014). Some interim conclusions have emerged from a World Bank review, which says that 'With the right combination of new policies and investments, countries can integrate unprecedented shares of variable renewable energy into their grids without compromising adequacy, reliability or affordability'. However, it warns that to best manage the challenge of integrating higher levels of variable renewable energy (VRE), 60% or more, into electricity grids, 'policy, planning and regulatory interventions should be designed to minimize overall system costs subject to meeting performance targets, rather than minimizing the costs of VRE generation alone' (Martinez and Hughes, 2015).

A similar point was made in a recent paper looking at high renewable energy mixes (30–50% and beyond) in the United States: it is the total system cost that matters, requiring judicious choice of balancing techniques and generation options (Becker et al., 2015).

DOI: 10.1057/9781137584434.0007

Within that general framework, there is a range of possible mixes, types and scales of renewable supply. There is a debate over whether the focus should be on using electricity options for heating, as currently envisaged. It may be hard to replace gas for heating, for example, in the United Kingdom, the gas grid carries about four times more energy (mostly for heating) than the electricity grid. Indeed gas grids might be better than electricity grids in some contexts, not least since gas can be stored. There is also a debate about what should be the right scale of renewables, large or small systems. All in all, there is plenty to discuss.

4.4 The renewables mix

As noted earlier, some renewables are best used on the medium to large scale, as with wind, wave and tidal farms, exploiting specific locationally defined energy resources. There are clear economies of scale for some renewables. For example, since the energy available from wind turbines is proportionate to the *square* of the radius length of the turbine blade and the *cube* of the wind speed, a large MW-scale machine on a windy hill will produce very much more output than the same total generation capacity of a series of small kW machines in low wind speed urban areas. Although PV solar can be used at any scale, there are advantages at larger scale, for example, reduced costs from bulk buying and a shared bulk installation/connection programme, with many projects being done at the same time, or with large solar farms being installed in easier to access and link up sites. Indeed MIT says power from utility-scaled PV can cost 70% less than from residential units, in the US context (MIT, 2015)

Enthusiasts for decentralisation argue, however, that smaller-scale localised systems offer social and political benefits, including opportunities for local control and ownership, as with solar energy. The late Hermann Scheer was a champion of this view, which he outlined in his 2005 Solar Manisfesto: 'Since everybody can actively take part, even on an individual basis, a solar strategy is 'open' in terms of public involvement... It will become possible to undermine the traditional energy system with highly efficient small-technology systems, and to launch a rebellion with thousands of individual steps that will evolve into a revolution of millions of individual steps' (Scheer, 2005). The prosumer and local energy co-op movement in Germany and elsewhere has now fleshed out this vision.

DOI: 10.1057/9781137584434.0007

At the extreme, enthusiasts for self-sufficiency look to totally off grid systems, and in some remote locations this may be the only option. However, in most parts of most industrialised countries, linking to the grid makes sense, since it allows users to export any excess electricity and top up from the grid when they are short. That is what the prosumers in Germany do, and it is central to the FiT system. It is also very helpful for overall balancing, so everyone gains, with local surpluses potentially being distributed widely and linking, via the grid, into outputs from larger more remote renewable projects.

However, new technological developments may alter the balance. PV solar is getting cheap and new battery storage systems are also getting cheap, so that some say consumers could run independently. In parts of the United States, for example, there is talk of 'grid defection' (RMI, 2014), and, given the spread of PV, the same issue could be relevant in Europe (Spross, 2014). A UK study noted that 'Until recently, individual storage units were not seen as a viable option, but prices have fallen rapidly (from $500/kWh in 2013, to $360/kWh in 2014) and financial institution, such as UBS, are predicting further cuts, with prices as low as $100/kWh within 10 years' (Mitchell et al., 2014).

It went on:

> This accelerated expected decline in storage costs is a reflection of the confidence in the development of batteries for Electric Vehicles (EVs). Batteries are now ubiquitous and there is clear cross-over between the technology developments in the different sectors. It was the drive for lower costs for laptop batteries that accelerated falling prices in the EV sector. However, now it is EVs that are driving down storage costs. The most graphic example is the Tesla company, one of the world's leading EV manufacturer, which announced in September 2014 that it will build a $5 billion 'Giga factory' that will double the global annual production of EV batteries and potentially half their production costs.

Tesla later launched $300/kWh home-scale battery units.

There is certainly enthusiasm for a more decentral approach, with PV and storage by consumers being seen as a key. However, not everyone is convinced that this individual-house approach, even if it also expanded to include PV on other buildings (shops, offices, factories warehouses), along with the necessary backup batteries, would be able to supply sufficient energy for all needs, year round. In particular it seems unwise and unrealistic to forgo the potential inputs to the grid from larger systems like wind, wave and tidal farms. It is also not clear if this independent

DOI: 10.1057/9781137584434.0007

approach would be optimal in energy terms, and in terms of balancing variations in energy output and demand (Borenstein, 2015). A key problem is that energy storage of whatever kind is still expensive, especially on the small scale, and as illustrated earlier, there are other, arguably better, ways to balance variable renewables, including using the grid. So there is a debate on the merits of and prospects for energy storage (POST, 2015).

4.5 The storage issue

It is certainly true that most forms of storage are at present expensive. Indeed this may always be the case, since, by definition, storage systems deliver energy only for part of the time, so their capital cost/kWh output is bound to be high. The only time that storage makes economic sense is when there is no alternative energy source available; that is why we pay a huge amount (in £/kWh terms) for small portable batteries, for example, for torches and radios. But in terms of bulk energy supply, storage is viable only if the cost of the input energy is very low and/or the price that can be charged for the energy output is high. That would be the case when variable renewables supplies are low, but energy demand high, and no other sources are available. However, in most cases, other sources *are* available, and are often low cost, such as natural gas. So the argument goes, storage will never compete with cheap gas turbines, which already exist on the grid. Especially if there is a need to store energy over any length of time – hours or days or even weeks. Gas can be stored for long periods, at low cost and with very low losses. Indeed gas pipeline systems act as an energy storage buffer.

This issue has come to a head in Germany, where a leading think tank has claimed that energy storage may not be cost effective in Germany in helping the transition to a higher grid penetration of renewable energy until that penetration level reached as high as 90%. Agora Energiewende's report, 'Electricity Storage in the German Energy Transition', said that, in the next 10–20 years, the use of energy storage was not likely to be more effective for allowing renewable energy to be integrated into the grid than a mixture of other options aimed at giving energy systems more added flexibility.

However, the Agora study did see circumstances where storage could help, for example, with fast, short-term grid balancing for use when the grid suddenly lost power due to a large plant failure or sudden demand

DOI: 10.1057/9781137584434.0007

rise. An Agora spokesman said: 'So far this can only be provided by fossil fuel power plants and what we saw in the study was that batteries can add value there and are really well technically suited for that'. And he did not dismiss domestic-scale batteries used with PV. Customer-sited storage could have a 'relieving effect on the distribution network when used in a grid-supporting manner'. The batteries in Electric Vehicles might also do that: the 'vehicle-to-grid' idea, with EV batteries being charged from the grid overnight at home but also being available to meet demand peaks on the grid if needed. But large-scale bulk storage was not needed yet (Colthorpe, 2014).

Agora's view flies in the face of much conventional thinking, and drew criticism from amongst others, DENA, a leading German Energy Agency. Stephan Kohler, its chief executive, said: 'Electricity storage facilities are essential for the energy turnaround. Anyone who alleges otherwise is damaging the energy turnaround and, in the end, is risking the supply security in Germany' (DENA, 2014b).

Certainly large-scale pumped hydro storage is often seen as a key way forward for balancing variable renewables, along with newly emerging ideas like Compressed Air Energy storage (Gaelectric, 2011) and Liquid Air Energy storage (Highview, 2015). Some flow batteries also look promising (Darling et al., 2014).

Some of this may get underway soon. A very ambitious $8 billion project has been proposed in the United States, involving a 2.1 GW wind farm in Wyoming sending electricity by HVDC grid 525 miles to an underground salt cavern compressed air storage facility in Utah, and then 490 miles on to Los Angeles (PennEnergy, 2014; Gruver and Brown, 2014). However, that is a one-off so far, and even the more established pumped storage option is facing challenges in the United States and the EU (Ela et al., 2013; Jones, 2012).

The debate on storage is surprisingly charged. Anti-renewables lobbyists sometimes argue that variable renewables need storage, and since that is expensive they are not viable. That may be why some pro-renewables lobbyists argue that large-scale storage is not necessary. And as can be seen, that may be true in general terms, at least not yet, for bulk storage, while there are still other plants on the grid, and may not be for some time, if grid balancing can be achieved by smart grid demand management and supergrid electricity exchanges. Although clearly, if cheap storage does become available, it could be very useful, possibly cutting system cost (Pfenninger and Keirstead, 2015).

DOI: 10.1057/9781137584434.0007

However, that depends on the context and the type: what is needed in the storage debate is more clarity about the scale and purpose of storage. Not all storage systems are the same. Some can store large amounts of energy for long periods (hours, days or even weeks), others are good at fast discharge for short periods. If you want very long-term inter-seasonal storage, heat stores are probably the best (e.g., for solar thermal), but if you want fast electric grid balancing, then conventional batteries or advanced chemical flow batteries are better. Horses for courses. Some can be used for local voltage support and others for bulk storage. And there are other factors. Some options are geographically determined (e.g., pumped hydro and underground compressed air stores), others can be sited anywhere, nearer end users (e.g., liquid air stores), but still possibly at large scale, including large Lithium-ion Batteries (Kanellos, 2014).

The energy storage industry is facing unprecedented innovation: new ideas are being tested out and niches may exist for many of them, including big batteries: a 5MW/5MWh project has opened in Germany and some see that as a way ahead (Hales, 2014). The United Kingdom is also looking at large (2 MW) battery stores (Sheffield University, 2014).

Meanwhile domestic-scale storage continues to make some inroads. It will be interesting to see which approach wins out. Batteries are certainly making progress, Lithium Ion units especially, for example, for domestic short-term electricity balancing (Normark et al., 2014; Tesla, 2015). However, the focus may not just be on batteries, or even just electrical power, given the need for longer-term bulk energy storage and for heat supply. For example, on the larger scale, there are large heat store systems linked to district heating networks in Denmark and elsewhere, fed by solar heat. They capture summer heat for use in winter. Excess electricity from wind farms could also be stored. A similar interseasonal solar-fed heat store system is used for a housing complex in (often very cold) Canada.

At the same time, on the small scale, there are systems enabling individual domestic consumers to feed any excess electricity from their rooftop PV array to their hot water boiler heat store, using an immersion heater. This means they can use less gas for heating, though direct solar water heating might actually be a cheaper option than using PV. Moreover, if you have excess PV electricity, it may be better to export it. Then others can use the electricity. Offsetting relatively low carbon directly used fossil gas in homes may not be the best interim option: it is

DOI: 10.1057/9781137584434.0007

better environmentally to offset high carbon fossil-derived electricity. As can be seen, the battle over scales, types, storage and sources continues!

Certainly the deployment of PV solar has led to some conflicts, with storage adding a new dimension. It is worth looking at this in more detail, given the implications for system integration. Domestic consumers with PV still usually use the grid system to earn extra income from exporting any excess and also for backup to meet shortfalls, sometimes without contributing to the overhead costs of running the grid system. That may be seen as unfair, and there are moves to charge prosumers' system-use charges. However, prosumer self-generation means that the grid system is relieved of the need to supply power some of the time, which should offset some of the extra local grid management/upgrade costs they impose by the need to manage the power they export. If prosumers invested in local domestic energy storage that would change the situation: some might even go off grid entirely. In that opt-out situation, though they would not be using the grid, the system costs for the rest would still rise, since there would be fewer linked-in consumers sharing the residual costs. So there is plenty of room for misgivings. If PV users stay on the grid, but do not pay a surcharge, they can be seen as 'free riders', enjoying an 'elite' private option paid for by the poor. Equally though some see attempts to get PV users to pay system costs as a way for the utilities to resist the spread of PV (and home storage) and the disruption it causes to their operations and profits. The utilities are clearly being challenged. Some of them even worry that consumer might store cheap off-peak grid power and sell it back during peak demand times at higher prices! A study by MIT noted that conflicts over these retail price issues have already emerged in the United States, and says 'robust, long-term growth in distributed solar generation likely will require the development of pricing systems that are widely viewed as fair and that lead to efficient network investment' (MIT, 2015). However, that will not be easy. Certainly that is indicated by the battles over a new tax on PV and storage in Spain (Pentland, 2015). Clearly, system design and integration involves more than just technology.

4.6 Conclusions

Integrating renewables into a viable energy supply and demand system can be done to some extent piecemeal, but the newly emerging system will be very different and there will need to be a holistic view to establish an optimal approach. At present, that is some way off, although recent studies offer some policy and economic insights on the impacts of various UK possible mixes.

The overall prognosis is good. As the study mentioned earlier of electricity supply mixes by researchers at Imperial College London concluded, 'even with the conservative cost assumptions used, achieving renewables shares above 80% is feasible from a cost perspective and from a technical perspective to the degree that hourly data can demonstrate this' (Pfenninger and Keirstead, 2015). In terms of grid balancing, and on the basis of detailed spacial (20 UK regions) and temporal (hourly) modeling, it saw storage as being important: 'The availability of grid-scale storage in scenarios with little dispatchable generation can reduce overall levelized electricity cost by up to 50%, depending on storage capacity cost.' But for high renewable scenarios, it also saw a need for supergrid links and firmer renewables: 'For more than an 80% renewable generation share to be economically feasible, large-scale storage, significantly more power imports, or domestic dispatchable renewables like tidal range must be available.'

Also, looking at the balancing technology choices, and using hourly modeling, a study from researchers at Southampton University specifically compared the cost of an interconnector-based approach with storage-based approaches in the United Kingdom: it said storing hydrogen in underground caverns offered the lowest cost long-term solution (Alexander et al., 2015). However, a study for Agora Energiewende, covering Germany, France, Switzerland, Austria and the Benelux countries, concluded that cross-border interconnection 'mitigates flexibility needs from increasing shares of wind and solar' and had many overall system integration advantages, reducing the need for storage and the level of surplus power curtailment (Agora, 2015). That contrasts strongly with the results of a more critical study (on wind) for the UK-based Adam Smith Institute, which claimed that, since at times there would be no wind power available to exchange via an EU supergrid, storage and/or massive backup would be vital (Aris, 2014). Clearly debate continues over the balancing options, and indeed, from a similarly somewhat contrarian perspective, as to whether the system as a whole could work reliably without (or perhaps even with) continued backup (Andrews, 2015).

Certainly there are interesting and important divergencies of views, even amongst those who strongly support a green energy transition,

DOI: 10.1057/9781137584434.0007

for example, over the right supply scales and mixes within the system. As noted earlier, some see large-scale systems linked via supergrids as an anathema, and more of the same, others fear that fully decentralised systems will only be part of the answer, and look to a mix of both small and large, so as to aid system balancing. Many environmentalists oppose large hydro, although, as indicated, it may be able to play a key role in balancing variable renewables. Some 'greens' also oppose the use of biomass. And yet biomass offers a storable fuel resource, which can be used to provide 'firm' power for grid balancing.

Some of the wider policy conflicts may be unresolvable, but some may be resolved or changed by technological developments. There is a flurry of innovation and rethinking on how to balance energy supply and demand. New views and new technologies are emerging across the board. For a fascinating overview see 'The Exergeia Report' (Martin, 2015), which includes nano-technology solar developments. For an equally breathtaking survey see Keith Barnham's 'The Burning Answer' (Barnham, 2014). That focuses on small-scale local solar for electricity and also perhaps syn-fuel production. Developments like this may change the whole energy equation, while opening up new integration issues and options.

At the much larger scale there have been proposals for installing large Concentrating Solar Power (CSP) plants or large PV solar arrays in desert areas of North Africa and transmitting some of the electricity back to the EU, via undersea HVDC supergrids. Similar ideas have been proposed for CSP in the Gobi desert, feeding supergrid links across Asia. In an early version of the European version of this 'Desertec' idea, as well as providing for local needs, 15% of the EUs electricity was to be generated in this way (Desertec, 2015). That would open up many issues concerning, for example, the terms of trade with mostly poor host countries, the risk of 'land grab' exercises by investors from the richer northern countries and also security of supply: as with oil, the EU would be dependent once again on energy imported from overseas. It could also be argued that EU countries can and should generate their own green power and that importing it might provide an excuse not to (Elliott, 2012).

Major regional initiatives like this may take time, and will need careful assessment. At present, the focus has moved away from supergrid links to Desertec-type projects, and, within the Middle East-North Africa region, on to CSP just for local uses, but the huge desert solar resource, and the potential for long-distance access to it, is unlikely to be ignored in the long term, and in the meantime, interconnectors and supergrid networks are beginning to spread piecemeal across the EU and elsewhere, to aid energy trading and renewable balancing. In the long term, given their potential for grid balancing, we are likely to see many more supergrid networks spread, linking up resources in, and across, many countries, and even across whole regions. Indeed it might even eventually be possible to have a global network, a global energy internet, allowing daytime solar to be used by those in nighttime locations (Chatzivasileiadis et al., 2013).

Ideas for global grid integration like this may seem utopian and certainly are some way off. They are also very technology focused, as has been most of the discussion earlier looking at the energy supply side, large and small. However, changes on the demand side are equally important. As already noted, there are ideas for dynamic demand management and local energy storage, which could mean that that demand peaks are adjusted to match supply availability, the reverse of the historical approach. At the same time there is a need to reduce overall energy use and energy waste, which will help reduce the balancing problem, since then fewer renewable supply inputs are needed to meet the reduced demand. Certainly there are many opportunities to do this in all sectors, with, for example, potential savings of up to 40% by 2030 having been identified in the United Kingdom (DECC, 2012). Energy use in buildings, in the domestic sector especially, has been a major focus and much more can clearly be done (IEA, 2013). Indeed, some say that such is the potential for energy saving that this should be the major focus, rather than supply (Olivier, 2012).

However, there may be limits to the success of energy saving projects as a means of reducing total demand, or even peak demand, including the so-called rebound effect. If consumers save money by investing in energy efficiency or smart metering, they may spend the money saved on other more energy intensives good and services, thus undermining at least some of the carbon emission savings. There has been a long running debate over the likely scale of this effect, but most agree that, although there may be ways in which it can be reduced, it could be significant (Druckman, 2011; Gillingham et al., 2014). There has also been a long running debate over whether operating on the demand side, by installing energy saving measures in houses, is as cost effective as an option as is sometimes assumed. For example, some claim that it is easier to insulate a pipe than a house, so linking to district heating networks can

DOI: 10.1057/9781137584434.0007

be cheaper than fully insulating old buildings (JRC, 2012; Elliott, 2015; ETI, 2015).

More generally it is sometimes argued that operating on the supply side can avoid having to make social changes and lifestyle adjustments, some of which may be strongly resisted: it is the classic technical fix, trying to avoid unwelcome social change. That issue is taken up in Chapter 6, where it is argued that both technical and social change are likely to be needed, along with political policy changes. As should be clear from the present chapter, major changes will also be needed to ensure that grid systems are integrated and balanced. Most of the specific changes are technical, but, as with the overall process of shifting to a sustainable energy future, they will also require political and institutional change at the national and global level: new policies for a new future. That is the focus of the next chapter.

4.7 References

Aboumahboub, T., Schaber, K.,Tzscheutschler, P. and Hamacher, T. (2010) 'Optimization of the Utilization of Renewable Energy Sources in the Electricity Sector', Recent Advances in Energy and Environment Conference: http://www.wseas.us/e-library/conferences/2010/Cambridge/EE/EE-29.pdf

Agora (2015) 'The European Power System in 2030 – Flexibility Challenges and Integration – Benefits', Fraunhofer Institute Report for Agora Energiewende: www.agora-energiewende.org/service/publikationen/publikation/pub-action/show/pub-title/the-european-power-system-in-2030-flexibility-challenges-and-integration-benefits/

Alexander, M., James, P. and Richardson, N. (2015) 'Energy Storage against Interconnection as a Balancing Mechanism for a 100% Renewable UK Electricity Grid', *IET Renewable Power Generation*, 9 (2), March, pp. 131–141: http://digital-library.theiet.org/content/journals/10.1049/iet-rpg.2014.0042

Andrews, R. (2015) 'The Difficulties of Powering the Modern World with Renewables', Energy Matters web site, 10 June: http://euanmearns.com/the-difficulties-of-powering-the-modern-world-with-renewables

Apt, J. and Jaramillo, P. (2014) 'Variable Renewable Energy and the Electricity Grid', Routledge, London: http://www.routledge.com/books/details/9780415733014/

Aris, C. (2014) 'Wind Power Reassessed: A Review of the UK Wind Resource for Electricity Generation', Scientific Alliance Report for the Adam Smith Institute, London: www.adamsmith.org/wp-content/uploads/2014/10/Assessment7.pdf

Barnham, K. (2014) 'The Burning Answer,' Weidenfeld and Nicolson, London: https://www.orionbooks.co.uk/books/detail.page?isbn=9781780225333

Becker, S., Frew, B., Andresen, G., Jacobson, M., Schramm, S. and Greiner, M. (2015) 'Renewable Build-Up Pathways for the US: Generation Costs Are Not System Costs', *Energy*, 81, 1 March, pp. 437–445: http://www.sciencedirect.com/science/article/pii/S0360544214014285

Borenstein, S (2015) 'Is the Future of Electricity Generation Really Distributed?', Energy Institute at Haas, University of California Berkeley Blog: https://energyathaas.wordpress.com/2015/05/04/is-the-future-of-electricity-generation-really-distributed/

Boyle, G. (ed) (2009) 'Renewable Electricity and the Grid', Earthscan, London: http://www.routledge.com/books/details/9781844077892/#description

Chatzivasileiadis, S., Ernst, D. and Andersson, G. (2013) 'The Global Grid', *Renewable Energy*, 57, September, pp. 372–383: http://www.sciencedirect.com/science/article/pii/S0960148113000700

Colthorpe, A. (2014) 'Report Challenges Short-Term Role of Storage in Germany's Energy Transition', PV Tech, 22 September: http://www.pv-tech.org/news/energy_storage_not_needed_in_germany_until_nation_hits_90_renewables_penetr

Darling, R., Gallagher, K., Kowalski, J., Ha, S. and Brushett, F. (2014) 'Pathways to Low-Cost Electrochemical Energy Storage: A Comparison of Aqueous and Nonaqueous Flow Batteries', *Energy Environmental Science*, 7, September, pp. 3459–3477: http://pubs.rsc.org/en/Content/ArticleLanding/2014/EE/C4EE02158D

DECC (2012) 'Capturing the Full Electricity Efficiency Potential of the UK', Department of Energy and Climate Change, London: http://webarchive.nationalarchives.gov.uk/20121217150421/http://www.decc.gov.uk/en/content/cms/emissions/edr/edr.aspx

DENA (2014a) 'Pilot- und Demonstrationsprojekte im Power-to-Gas-Konzept', Deutsche Energie-Agentur: http://www.powertogas.info/power-to-gas/interaktive-projektkarte.html

DOI: 10.1057/9781137584434.0007

DENA (2014b) 'Dena Calls for Rapid Expansion of Electricity Storage Facilities,' DENA Press Release, 7 October: http://www.dena.de/en/press-releases/pressemitteilungen/dena-fordert-stromspeicher-muessen-zuegig-ausgebaut-werden.html

Desertec (2015) Desertec Foundation web site: http://www.desertec.org/concept/

Druckman, A., Chitnis, M., Sorrell, S. and Jackson, T. (2011) 'Missing Carbon Reductions? Exploring Rebound and Backfire Effects in UK Households', *Energy Policy*, 39 (6): http://www.sciencedirect.com/science/article/pii/S0301421511002473

Ela, E., Kirby, B., Botterud, A., Milostan, C., Krad , I. and Koritarov, V. (2013) 'The Role of Pumped Storage Hydro Resources in Electricity Markets and System Operation', US National Renewable Energy Laboratory paper: http://www.consultkirby.com/files/NREL-CP-5500-58655_Role_of_Pumped_Storage.pdf

Elliott, D. (2012) 'Emergence of European Supergrids,' *Energy Strategy Reviews*, 1 (3), March, pp. 171–173: http://www.sciencedirect.com/science/article/pii/S2211467X12000120

Elliott, D. (2015) 'Energy Use in Buildings', *International Journal of Ambient Energy*, 36 (2) February, p. 49: http://www.tandfonline.com/doi/full/10.1080/01430750.2015.1013007#abstract

ETI (2015) 'Decarbonising Heat for UK Homes', Energy Technologies Institute, Loughborough and Birmingham: http://theeti.cmail20.com/t/j-l-ddjhol-otiukhuju-k/

Gaelectric (2011) 'CAES Compressed Air Energy Storage', Gaelectric web site: http://www.gaelectric.ie/index.php/energy-storage/

Gillingham, K., Rapson, D. and Wagner, G. (2014) 'The Rebound Effect and Energy Efficiency Policy', E2e Working Paper 013, University of California, Berkeley, Massachusetts Institute of Technology and the University of Chicago: http://e2e.haas.berkeley.edu/pdf/workingpapers/WP013.pdf

Gruver, M. and Brown, M. (2014) 'eRenewable Energy Plan Hinges on Huge Utah Caverns', Associated Press, 24 September 24: http://bigstory.ap.org/article/3084cb4c459f4ffd9b666f5d5d2e44e3/wind-energy-proposal-would-light-los-angeles-hom

Hales, R. (2014) 'The First 100% Green Grid Is Online, Figuratively Speaking', Cleantechnica, 16 September: http://cleantechnica.com/2014/09/16/fist-100-green-grid-online-figuratively-speaking

DOI: 10.1057/9781137584434.0007

Highview (2015) High View Power Storage web site: http://www.highview-power.com/
IEA (2013) 'Energy Efficient Building Envelopes', Technology Road Map, International Energy Agency, Paris: http://www.iea.org/publications/freepublications/publication/TechnologyRoadmapEnergyEfficientBuildingEnvelopes.pdf
IEA (2014) 'The Power of Transformation – Wind, Sun and the Economics of Flexible Power Systems', International Energy Agency, Paris: http://www.iea.org/textbase/npsum/givar2014sum.pdf press release
IEA (2015) 'How2Guide for Smart Grids in Distribution Networks', International Energy Agency, Paris: www.iea.org/publications/freepublications/publication/TechnologyRoadmapHow2GuideforSmartGridsinDistributionNetworks.pdf
IPCC (2014) 'Fifth Assessment Report', Intergovernmental Panel on Climate Change, Geneva: http://www.ipcc.ch/
Jones, L. (ed) (2014) 'Renewable Energy Integration', Elsevier, London: http://www.elsevier.com/books/renewable-energy-integration/jones/978-0-12-407910-6
Jones, S. (2012) 'Mountain Ahead for "Battery of Europe"', Utility Week, 11 June: http://www.utilityweek.co.uk/news/mountain-ahead-for-battery-of-europe/824882#.VRQVgYUhuRp
JRC (2012) 'District Heating and Cooling', European Commission Joint Research Centre, Petten: http://setis.ec.europa.eu/system/files/JRCDistrictheatingandcooling.pdf
JRC (2013) 'Assessment of the European potential for PHS', European Commission Joint Research Centre: http://setis.ec.europa.eu/newsroom-items-folder/jrc-report-european-potential-pumped-hydropower-energy-storage
Kanellos, M. (2014) 'Will Lithium Ion Work for Grid-Scale Storage?', Renewable Energy World, 2 October: http://www.renewableenergyworld.com/rea/blog/post/2014/10/will-lithium-ion-work-for-grid-scale-storage
Martin, M. (2015) 'Energy Transition Fast Forward! Scouting the Solutions for the 80–100% Renewable Economy: The Exergeia Report', Impact Economy, Geneva: http://www.impacteconomy.com/en/primer4_details.php

DOI: 10.1057/9781137584434.0007

Martinez, S. and Hughes, W. (2015) 'Bringing Variable Renewable Energy Up to Scale: Options for Grid Integration Using Natural Gas and Energy Storage', World Bank Report: http://documents.worldbank.org/curated/en/2015/02/24141471/bringing-variable-renewable-energy-up-scale-options-grid-integration-using-natural-gas-energy-storage

MIT (2015) 'The Future of Solar Energy', Massachusetts Institute of Technology, Cambridge, MA: http://mitei.mit.edu/futureofsolar

Mitchell, C., Froggatt, A. and Hoggett, R. (2014) 'Governance and Disruptive Energy System Change', Conference Paper, International Workshop on Incumbent – Challenger Interactions in Energy Transitions', 22–23 September, University of Stuttgart, Germany: http://projects.exeter.ac.uk/igov/wp-content/uploads/2014/09/Post-stuttgart-1-final-paper.pdf

Normark, B., Faure, A., Deane, P. and Pye, S (2014) 'How Can Batteries Support the EU Electricity Network?', Insight-Energy Report for the European Commission: http://www.insightenergy.org/featured_topics?page=2#featured-topic-2

Olivier, D. (2012) 'Less Is More', Association for Environment Conscious Building: http://www.aecb.net/publications/less-is-more-how-we-can-keep-going-without-breaking-the-planet-or-the-bank/

PennEnergy (2014) '$8B Renewable Energy Initiative Proposed for Los Angeles', PennEnergy, 23 September: http://www.pennenergy.com/articles/pennenergy/2014/09/8b-renewable-energy-initiative-proposed-for-los-angeles.html

Pentland, W. (2015) 'Energy Storage Is the Real Target of Spain's New Tax on the Sun', Forbes,18 June: http://www.forbes.com/sites/williampentland/2015/06/18/energy-storage-is-the-real-target-of-spains-new-tax-on-the-sun/

Pfenninger, S. and Keirstead, J. (2015) 'Renewables, Nuclear, or Fossil Fuels? Scenarios for Great Britain's Power System Considering Costs, Emissions and Energy Security', *Applied Energy*, 152, 15 August, pp. 83–93: http://www.sciencedirect.com/science/article/pii/S0306261915005656

POST (2015) 'Energy Storage', UK Parliamentary Office of Science and Technology, POSTNote, 492: http://www.parliament.uk/briefing-papers/POST-PN-492/energy-storage

Poyry (2011) 'Analysing Technical Constraints on Renewable Generation to 2050', Pyory Consultants Report to the Committee on Climate

DOI: 10.1057/9781137584434.0007

Change, March: http://www.poyry.com/sites/default/files/technicalconstraintsonrenewablegeneration-march2011.pdf

Pugwash (2013) 'Pathways to 2050: Three Possible UK Energy Strategies', British Pugwash Group, London: http://britishpugwash.org/pathways-to-2050-three-possible-uk-energy-strategies/

RMI (2014) 'The Economics of Grid Defection,' Rocky Mountain Institute, Colorado: http://www.rmi.org/electricity_grid_defection - economics_of_grid_defection

Rodriguez, R., Sarah Becker, S., Andresen, G., Heide, D. and Greiner, M. (2014) 'Transmission Needs across a Fully Renewable European Power System', *Renewable Energy*, 63, March, pp. 467–476: http://www.sciencedirect.com/science/article/pii/S0960148113005351

Scheer, H. (2005) 'A Solar Manifesto', Earthscan, London: http://www.routledge.com/books/details/9781902916514/

Sheffield University (2014) 'Giant Battery to Help Tackle Energy Storage Challenges', University of Sheffield Press Release, 24 July: http://www.sheffield.ac.uk/faculty/engineering/enews/giant-battery-1.426169

Sorensen, B. (2014) 'Energy Intermittency', Routledge, London: http://www.routledge.com/books/details/9781466516069/

Spross, J. (2014) 'Home Solar Plus a Battery Could Be Cheaper Than the Grid in Germany in Just a Few Years' Climate Progress', 3 October: http://thinkprogress.org/climate/2014/10/03/3575371/hsbc-solar-battery-germany/

Tesla (2015) 'Powerwall Lithium-Ion 7/10kWh Domestic Batteries', Tesla Press Release: http://www.teslamotors.com/powerwall

DOI: 10.1057/9781137584434.0007

5

Policy Issues: How to Change the World

Abstract: *There are many scenarios suggesting that it would be possible to move to energy systems with near 100% of electricity, and perhaps of total energy, being supplied from renewable energy sources by around 2050. This may not be possible in every country, but there are some ambitious plans for moving in that direction. Most focus on wind and PV solar, along with biomass, although for some environmentalists the latter remains controversial if on a large scale, as does large hydro. However, there is more than enough potential to allow for a range of mixes, with a key issue being the appropriate balance of large and small-scale systems and the overall direction of travel, for example, to what extent are social changes also needed?*

Keywords: appropriate scales; energy policy; renewable acceleration

Elliott, David. *Green Energy Futures: A Big Change for the Good.* Basingstoke: Palgrave Macmillan, 2015.
DOI: 10.1057/9781137584434.0008.

 DOI: 10.1057/9781137584434.0008

5.1 The way ahead

There are now many scenarios outlining possible pathways ahead for supplying near 100% of global electricity or even of energy from renewable sources by around 2050, most of them taking balancing needs into account (Cochran et al., 2014). They include the already-mentioned '100% by 2050' scenarios for the EU, and a US study with 80% and perhaps up to 90% of electricity coming from renewables by 2050 (Mai et al., 2012). In addition, new high renewable scenarios, backed by energy saving, have emerged for many other key countries and regions, including China, India, South Korea, Japan (ISEP, 2013; WWF, 2013, 2014; Sorensen, 2014). How can they be turned into reality?

It is unlikely to be sufficient just to leave it up to market forces. Renewables may be getting cheaper, and cost-cutting energy saving may be making even more sense as conventional energy prices rise, but there are still powerful forces defending and promoting the existing supply and energy use energy options. They are being challenged, often on a case-by-case basis, by environmental and other grass roots groups, but to move ahead more rapidly, governments also have to play a role, at the very least by resetting markets and imposing emission controls. The rest of us also have a role to play, in terms of changing the way we use energy. Efforts to 'nudge' people into making changes may have some impact, and some behavioural changes may well be both good for us and welcomed, but few will take kindly to imposed changes in behaviour. However, changes will be imposed involuntary if and when the climate situation worsens. It remains to be seen to what extent significant behavioural change will emerge before then: demand for electricity is falling in some sectors and countries, but as is explored in Chapter 6, some may find it hard to escape the lure of consumerism and growth. So the appeal of supply-side technical fixes may be strong.

The current state of play on the energy supply side is certainly quite promising. As illustrated earlier, renewables are expanding rapidly, supplying 19% of total global energy by 2014, driven by a mixture of technological advance, government initiatives and, in some locations, prosumer enthusiasm, and progress is also being made on taming demand. But globally this is being offset by increases in the use of fossil fuels, partly in response to growing energy demand in some parts of the world, but also since some fossil fuel prices have been (temporarily) lowered by the advent of cheap shale gas, chiefly in the United States.

DOI: 10.1057/9781137584434.0008

However, there are signs of progress. As noted earlier, the IEA says that emissions from the energy sector in 2014 were at the same level as in 2013, which, given that world's economies had in general moved out of recession, suggests that policy responses, rather than economic factors, were having an impact (Briggs, 2015). More robust policies to reduce the emphasis on fossil fuel may help that trend to continue, but only if new non-fossil energy sources are developed. That had been the focus of this book.

The leaders in that regard are China, which now has 115 GW of wind capacity installed, and Germany, with around 80 GW of wind and PV solar in place, and aiming to get over 80% of its electricity from renewables by 2050. The United States gets around 15% of its electricity from renewables at present, hydro included, and could well expand that dramatically, as new renewables develop. The US Department of Energy says that wind could supply*10% of US electricity by 2020, 20% by 2030 and 35% by 2050 (DoE, 2015).*

While Germany may be the technological leader within Europe, the EU as a whole aims to get 20% of its total energy from renewables by 2020 and 27% by 2030, while cutting energy use by 27% by then (EC, 2014). Within the overall EU targets, some countries have been given very high sub-targets. Sweden has made the most progress so far. It has already reached 52%, overtaking its 48% 2020 renewable energy target. Several other EU countries are nearing their high 2020 renewable energy targets, although there are some laggards. They include the United Kingdom, which, by 2014, had only got to 5.1%, well short of its quite low 15% by 2020 renewable energy target (EurObsersvER, 2014).

In common with some other EU countries, the United Kingdom has found it harder to make as much progress in the heating and transport sectors as in the electricity sector (where it has reached 19% by 2013). Its strategy is to try to decarbonise heating and transport by using low-carbon electricity, from renewables and (if its new progamme gets going) nuclear, possibly along with shale gas with CCS sequestration. That electricity would be used to power a new fleet of electric vehicles, and on the heat side, would replace the use of gas for home heating, along with some biomass, including CHP-fed networks. Solar-fed district heating systems with heat stores might also be a candidate, as is being followed up, for example, in Denmark. However, UK progress with heat pump deployment to replace gas use and electric vehicles to replace petrol and diesel use has been slow and there are disagreements about strategy. For example, some think biogas would be a better vehicle fuel and (district)

DOI: 10.1057/9781137584434.0008

heating option, while others look to synfuels and hydrogen, perhaps used with fuel cells in vehicles.

Can progress in the United Kingdom and elsewhere be improved, extended and speeded up? The main constraint is not so much technology, although technical advances that reduce costs will help. The main problem, looking broadly, is the lack of political will. China has made commitments to constrain carbon emission growth by 2030, the United States has set a target of 26–28% reduction by 2025 and the EU aims to reduce emissions by perhaps 40% by 2030, on condition that other countries too make strong commitments. These plans are welcome, but may not be enough to avoid the risk of major climate change impacts.

5.2 Accelerated expansion

How could we do better? The potential is certainly there. In terms of technology, wind power could reach a total installed global capacity of up to 2,000 GW (2TW) by 2030, supplying up to 19% of global electricity. That is a key conclusion of the latest Global Wind Energy Outlook from the Global Wind Energy Council and Greenpeace. That is in the advanced scenario. A more moderate scenario puts it at just under 1,500 GW by 2030 (Appleyard, 2015). PV solar could also make a major contribution. One quite conservative study suggested that Germany, a far from sunny country, could ultimately have 275 GW of PV, while India has recently set a target of 100 GW of PV by 2022 (ET, 2015).

In terms of national programmes, the International Renewable Energy Agency's report on the United States, in its REMap 2030 series, says it can increase the use of renewables in its energy mix to 27% by 2030 (the same as the EU's target), and in the power sector alone to almost 50% (IRENA, 2015a). Its study of China suggests that it could use renewables to supply 26% of its total energy by 2030, up from 13% now, and 40% of its electricity by then, up from around 20% now. That would involve 400 GW of hydro, 560 GW of wind and 308 GW of PV. For heat, solar and (more problematically) biomass expand significantly. Coal use levels off by around 2020, but gas and oil use expand slightly. So does nuclear, but less so (IRENA, 2014a).

IRENA looks to modern biomass as a major option, possibly supplying 20% of global energy by 2030. It suggests that around 40% of the total global biomass supply could come from agricultural residues and

DOI: 10.1057/9781137584434.0008

waste, and another 30% from sustainable forestry products. IRENA says these biomass sources would not compete with the resources needed for food production, such as land and water (IRENA, 2014b).

Some studies suggest that the levelised cost of bioenergy could be reduced by almost half by 2025, making bio-based electricity close to competitive with coal. With demand for biofuels also rising, there may be strong pressures to expand biomass use (Albani et al., 2014). IRENA says: 'Sustainably sourced biomass, such as residues, and the use of more efficient technology and processes can shift biomass energy production from traditional to modern and sustainable forms, simultaneously reducing air pollution and saving lives.' However, there are counterviews, and it is worth exploring the biomass issue in more detail.

At present biomass provides about 10–15% of the global total primary energy supply, of which 60% is used in traditional households mostly in developing countries, some 25% for heat and power generation, largely in developed countries (around 90 GW), the rest being used in informal sectors such as charcoal and brick making, almost entirely in developing countries. The very inefficient way biomass has been used traditionally results in pollution and health problems and diminishing firewood and dung resources, but new technologies can, it is hoped, change that.

However, as has already been noted, there are environmental objections to the use of some types of biomass, in particular liquid biofuels for vehicle use, especially from plantations in developing countries (EP, 2012), but also stem-wood burnt as imported wood pellets in large power plants. The optimists claim that higher-yield second-generation non-food biofuel sources can help avoid the food versus fuel issue, make more productive use of land and avoid over exploitation, if coupled with careful regulation. Similarly, optimists say sustainable source regulation can ensure that stem-wood use is avoided for power production and suggest that the energy and carbon costs of importing biomass, even involving long-distance transport, are low, in which case large new markets could open up for producers in the developing world, with Biomass Energy Carbon Capture and Storage possibly adding a 'negative carbon' bonus.

Biofuels and electricity are high value products, so not everyone is convinced by these arguments: there are huge market forces which may over-ride sustainability concerns. At the very least, much tougher controls will be needed and most environmentalists would prefer a focus on smaller-scale systems, using less aggressive technology, for example,

DOI: 10.1057/9781137584434.0008

local biogas production from anaerobic digestion of farm and other biowastes. Biogas can then be used for heating, power and in vehicles.

Some, however, also look to growing special energy crops, although done on a sustainable basis. The European Forestry Institute (EFI) notes that 'more than 10 million ha. of set-aside fields are presently available in the EU for the cultivation of dedicated biomass tree crops', and suggested that this 'needs appropriate policies that reward short rotation tree cultivation for bioenergy and reduce uncertainties that deter the private sector from investing in new technologies'. The emphasis is thus on local production, with fewer imports. As the EFI notes, 'an important objective of the European bioenergy policy has been the decentralisation of renewable energy production leading to the increased utilisation of local energy sources, improved local energy security, shorter transport distances and lowered transmission losses' (Pelkonen et al., 2014).

The debate over biomass, including BECCS, continues. As indicated earlier, although some think it should not be used at all, much of the debate is on how it can be used and how problems can be overcome (Huenteler et al., 2014).

5.3 Putting it all together – constraints and opportunities

The details of specific renewable energy programmes obviously need to be thrashed out, on biomass especially, but the general picture is clear. Even if biomass is excluded, there is the potential for rapid expansion on the way to near 100% contributions by 2050 (Delucchi and Jacobson, 2013). Getting to 100% in energy as well as electricity by 2050 may not be possible in many countries, though Denmark is aiming for that, with wind playing a major role and biomass limited, and a UBA study says Germany could too, even without biomass (UBA, 2014). A Greenpeace Energy [R]evolution report suggested renewables could supply 92% of China's electricity by 2050, from over 3000 GW, including hydro and biomass. A new WWF study is more cautious, aiming at 80% of electricity by 2050 (WWF, 2014). But its India scenario aims for 90% of total *energy* by 2050 (WWF, 2013).

Looking globally, in the World Energy Council's 2050 global energy market-led 'Jazz' scenario, the share of renewables in electricity generation is 31% and in its more policy-led 'Symphony' scenario, 48% (WEC,

DOI: 10.1057/9781137584434.0008

2013). As we have seen, these may be conservative estimates. A major Global Energy Assessment, produced in 2012, noted that the share of renewable energy in global primary energy could increase 'to between 30% to 75%, and in some regions exceed 90%, by 2050' (GEA, 2012). Renewables have moved on dramatically since 2012, so perhaps this too is conservative, at least for some countries.

Some worry that there will be resource constraints on expansion on this scale, with the emphasis often put on the large material requirements, for example, for steel and aluminum, as well as more conventional concerns about impacts. However, a full life-cycle resource analysis has suggested that renewables could supply the world's entire electricity needs by mid-century without major problems with resource (materials) use or eco-impacts. It assessed the whole-life costs of solar, wind, hydro as well as gas and coal generators with carbon capture and storage. But it left out biomass as being too complex to assess. It looked at the demand for aluminum, copper, nickel and steel, metallurgical grade silicon, flat glass, zinc and clinker and the impact of greenhouse gases, particulate matter, toxicity in ecosystems and eutrophication (overwhelming plankton bloom) of the rivers and lakes. It found that to generate new sources of power, demand for iron and steel might increase by only 10%. PV solar systems would require between 11 and 40 times more copper than needed for conventional generators. But even so, 'only two years of current global copper and one year of iron will suffice to build a low-carbon energy system capable of supplying the world's electricity needs by 2050'.

The overall conclusion was that

> The large-scale implementation of wind, PV, and CSP has the potential to reduce pollution-related environmental impacts of electricity production, such as GHG emissions, freshwater eco-toxicity, eutrophication, and particulate-matter exposure. The pollution caused by higher material requirements of these technologies is small compared with the direct emissions of fossil fuel fired power plants. Bulk material requirements appear manageable but not negligible compared with the current production rates for these materials. Copper is the only material covered in our analysis for which supply may be a concern. (Hertwich et al., 2014)

Some fear that the diminishing fossil sources cannot support a transition to renewables, since renewables 'currently require fossil fuels for their construction and deployment, so in effect they are functioning as a parasite on the back of the older energy infrastructure. The question is, can they survive the death of their host?' (Heinberg, 2015).

DOI: 10.1057/9781137584434.0008

It certainly could be argued that we should be reserving as much of whatever fossil energy is left as possible to support this process, rather than just burning it off for no long-term gain, although that means there will be an emissions debt. It is also true that there are important non-energy uses for fossil resources (e.g., for around 13% of oil), so some should be reserved for that too (Kroll, 2013). However, to some extent, the first wave of renewables can provide energy to support the next wave, so perhaps this is not a major problem, depending on how quickly the transition needs to made, especially given that, as noted earlier (in Section 2.5), the Energy Returns on Energy Invested for renewables are higher and rising while those for fossil fuel are now low and falling and similarly for nuclear.

There are some regions where rapid expansion may be harder, in Africa, for example, due to local economic and political constraints, but that is a major current focus, for example, of the UN's Sustainable Energy for All programme, backed by the EU. Solar and biomass are obvious options. IRENA says that Africa has the potential and the ability to utilise its renewable resources to fuel the majority of its future growth with renewable energy. It adds 'doing so would be economically competitive with other solutions, would unlock economies of scale, and would offer substantial benefits in terms of equitable development, local value creation, energy security, and environmental sustainability' (IRENA, 2013). For an update, see the special issue of *Energy Research and Social Science*, 'Renewable Energy in Sub-Saharan Africa' (Hancock, 2015).

Renewables are doing quite well in South America. As in Africa, some countries already get the majority of their electricity from hydro and there are some impressive national programmes for expanding new renewables. For example, there is around 22 GW of PV solar capacity under development, with Chile, Brazil and Mexico in the lead, while wind is also expanding. Mexico has a 40 GW wind resource and may become the world's one of the fastest growing wind energy producers.

Clearly some poorer countries will face problems, but the IEA says that, for newly emerging economies, with less sunk costs in conventional energy systems, the transition costs and difficulties of switching to renewables would be less: they could 'leapfrog' to a new cleaner, greener and more efficient energy system (IEA, 2014).

In geopolitical terms, what happens in the Middle East may be seen as crucial. Saudi Arabia has an ambitious $109 billion clean energy programme, with an initial target for installing 41 GW of solar (25 GW CSP, 16 GW PV) by 2032. That has now been put back to 2040, but solar

DOI: 10.1057/9781137584434.0008

PV and CSP is making progress across the Middle East, with major projects in the UAE and elsewhere, as well as in North Africa. A key point is that solar is well matched to the regions growing daytime air conditioning load and desalination needs. For those with oil, it would also avoid diverting oil supplies for domestic power generation during the hottest summer months, which would reduce crucial export income. So for them there is a strong economic incentive (Lahn and Stevens, 2011).

Given its current government, it does not seem likely that Australia will make radical changes, despite the climate shocks it has experienced, and despite being ideally suited to solar development (BZE, 2010). Sadly, Canada seems to be in a somewhat similar state, and has a huge as yet only partly untapped wind potential. Russia is similarly intransigent: it is focusing on exporting its fossil reserves and developing nuclear. This despite the fact that it has huge wind resources in the vast wind-swept reaches of Siberia, up to 350 GWs worth. It has some large hydro projects, but its current target is only to get 4.5% of its electricity from new renewables by 2020.

Perhaps, symbolically, given its tragic experience with nuclear power, the situation in Japan deserves special mention. After the Fukushima nuclear accident in 2011, a major renewable energy programme was put in place, including the development of offshore wind, using floating systems. The first 2 MW floating device was installed off the Fukushima coast and many more are planned in a 1.45 GW programme. The long-term potential for offshore wind is put at 100 GW. PV solar is also being strongly supported via a quite generous Feed-In Tariff. Interestingly, though rooftop sites abound, as in India, given the land scarcity, floating PV arrays are being deployed on reservoirs. Although land-use constraints are an issue, the longer-term potential is still substantial. A '100% by 2050' ISEP renewables scenario has around 50 GW of wind, much of it offshore, and 140 GW of PV (ISEP, 2013). The near 100% scenario for Japan developed by Bent Sorensen moves the debate on further (Sorensen, 2014). He also offers 100% scenarios for South Korea, India, Mexico, China, Canada and the United States, all with grid balancing needs carefully assessed. In the case of Japan, current outline plans only envisage renewables expanding to supply around 24% of electricity by 2030, although details of the overall likely mix are scarce. The contribution from nuclear remains uncertain, with all the nuclear plants still

DOI: 10.1057/9781137584434.0008

closed. However, given its traditional technical ingenuity, it may be that Japan will be a pioneer in developing new cleaner and safer technologies. It certainly has the incentive.

5.4 Conclusions

The incentive for change will hopefully not have to be provided by nuclear disasters or climate threats or air pollution crises. It ought to be possible to move proactively to avoid major problems. That of course assumes there is a rational energy policy and policymaking system. The reality, in many countries, seems to be otherwise. Policy is often captured by powerful vested interests, able to defend and maintain the energy status quo, at least until the problems, and public reactions to them, reach a level when some sort of response cannot be avoided. At that point the risk is that panicky ill-conceived measures may be considered. For example, at present, desperate measures are being proposed to try to reduce climate change via large-scale planetary geo-engineering projects, which may be both risky and ineffective, while diverting resources from developing more sustainable solutions (Keller et al., 2014). Attempting to reduce the planet's solar heat gain by using giant reflector screens in orbit in space or aerosol particles injected into the upper atmosphere seem to be exactly the inverse of what should be done: the solar energy input should be used, not blocked.

As shown, the technical potential for doing this is very large. So is the opportunity for change. Globally there is very strong public support for renewables: they are popular. In a global opinion survey carried out by IPSOS across 24 countries just after Fukushima, solar power was backed by 97% of the sample, wind power by 93%, hydro power by 91%, natural gas by 80%, coal by 48% and nuclear by just 38% (IPSOS, 2011). Since then renewables have expanded significantly, and support for them continues to be high. For example, polling for the UK Department of Energy and Climate Change found that, in mid-2014, 36% of UK adults asked backed nuclear, whereas 79% backed renewables, offshore wind 72%, biomass 60%, onshore wind 67%, wave and tidal 73% and solar 82% (DECC, 2014).

As was noted in Chapter 1, some of the major oil companies are beginning to worry about their future, although they still see fossil

DOI: 10.1057/9781137584434.0008

fuel as staying central, and, perhaps inevitably, as their core business focus (Dargaville et al., 2015). However, some power engineering and energy supply companies have shifted their position, notable in Germany, with Siemens, RWE and most recently E.ON, moving away from fossil fuel and nuclear and on to renewables. In part that has been forced on them by energy market changes in Germany, resulting from new government policies and local prosumer initiatives. For example, given the power market changes, RWE has indicated that in future, rather than being just an energy supplier, it aims to service the new market for renewables and distributed energy and support system integration (Parkinson, 2013, 2014), this from a company whose then chief executive once said, 'Photovoltaics in Germany make about as much sense as growing pineapples in Alaska' (Steitz and Eckert, 2012).

Moreover the changes in perspective are wider-spread than just in Germany. The business case for moving into the new expanding area of renewables, and to gain 'first mover' advantages, is becoming clear across the world, with China leading. Elsewhere, making the change may involve breaking out of old patterns and commitments. But with the market shifting, this may not be more than good business sense. As an E.ON executive put it when it withdrew from the Horizon nuclear project in the United Kingdom, 'we have come to the conclusion that investments in renewable energies, decentralised generation and energy efficiency are more attractive, both for us and for our British customers' (Teyssen, 2012).

The debate over which way to go, and over which renewables to back most, continues in the United Kingdom and elsewhere, but as argued, whatever the outcome, much more will have to be done if we are to avoid climate crises and air pollution problems. There are debates on which specific mechanisms to use to accelerate the deployment of the necessary technologies and slow the use of inappropriate technologies. Some look to market-based mechanisms, like carbon emission permit trading, which penalise fossil fuel use and therefore advance non-fossil options. So far that has not been very successful. It has proved hard to get political agreement on carbon caps that were low enough to create an effective carbon market. That may change, but as we have seen, so far, direct subsidies, as with the Feed-In Tariff system, have been far more effective at stimulating the take-up of renewable energy technologies. FiTs are a

DOI: 10.1057/9781137584434.0008

form of indirect tax. It may be that in future direct carbon taxes will be introduced and accepted, if and when concerns about climate change rise, and meanwhile a gradual process of setting tighter emission limits is underway.

However, there is still resistance from some countries to setting tight carbon limits, given concerns about the alleged impacts on economies. In reality, this may be short-sighted. There may be short-term gains from continuing with dirty technologies, but those who delay making the transition to clean energy will be uncompetitive as emission controls are gradually tightened up across the world. There is also a broader historical perspective.

The first wave of industrialisation in the West, fuelled increasingly by fossil energy, enhanced wealth creation and also led to horrendous social and environmental problems, dangerous and unsustainable employment and arguably to a global economy trapped in growth at all costs (Elliott, 2015). The new wave of post-carbon technological and economic development will hopefully avoid, and learn from, these problems. So there ought be an incentive to make the change. Some see the creation of sustainable jobs as a positive incentive: at present around 7.7 million people are employed in the renewables sector globally and that is clearly set to expand (IRENA, 2015b). In addition, renewable energy development offers prospects for local economic and social renewal and the engagement of local communities in energy issues, including locally owned projects and new forms of social enterprise (Kunze and Becker, 2014).

Developments like this, along with market pressures and government initiatives, may create a new impetus for expansion. But there remains the issue of whether economic growth is the right aim. The final chapter looks more broadly at the issue of growth, and at the role that new technologies might play within a more sustainable global economy. A key question is whether renewables are just a technical fix to sustain an unchanged society, when in fact what is required is radical social change. While some see the adoption of renewables as part of a process of social change, and some social change as actually a prerequisite for wide-scale renewable expansion, a more radical view is that adoption renewables are not likely to be very helpful, and indeed could be counter-productive, unless there are also major social and economic changes (Huesemann and Huesemann, 2011).

DOI: 10.1057/9781137584434.0008

5.5 References

Albani, M., Denis, N. and Granskog, A. (2014) 'Can Bioenergy Replace Coal?', McKinsey & Company Consultants, September: http://www.mckinsey.com/insights/sustainability/can_bioenergy_replace_coal

Appleyard, D. (2015) 'Wind Energy Outlook 2015: Could Total Installed Wind Capacity Reach 2,000 GW by 2030?', Renewable Energy World, 4 February: www.renewableenergyworld.com/rea/news/article/2015/02/wind-energy-outlook-2015-could-total-installed-wind-capacity-reach-2000-gw-by-2030

Briggs, H. (2015) 'Global CO2 Emissions "Stalled" in 2014', BBC Report on IEA Press Story, 13 March: http://www.bbc.co.uk/news/science-environment-31872460

BZE (2010) Zero Carbon, Australia: http://beyondzeroemissions.org/zero-carbon-australia-2020

Cochran, J., Mai, T. and Bazilian, M. (2014) 'Meta-analysis of High Penetration Renewable Energy Scenarios', *Renewable and Sustainable Energy Reviews*, 29, January, pp. 246–253: http://www.sciencedirect.com/science/article/pii/S1364032113006291 - cor1

Dargaville, R., Workman, A., Lafleur, D., McConnell, D., Wang, C., Wainstein, M. and Alexander, R. (2015) 'BP's Extreme Climate Forecast Puts Energy Giant in a Bind', *The Conversation*, 31 March: http://theconversation.com/bps-extreme-climate-forecast-puts-energy-giant-in-a-bind-39250

DECC (2014) 'Public Attitudes Tracker: Wave 10', Department of Energy and Climate Change, London: https://www.gov.uk/government/statistics/public-attitudes-tracking-survey-wave-10

Delucchi, M. and Jacobson, M. (2013) 'Meeting the World's Energy Needs Entirely with Wind, Water, and Solar Power', *Bulletin of the Atomic Scientists,* 69 (4), pp. 30–40: http://bos.sagepub.com/content/69/4/30.full

DoE (2015) 'Wind Vision: A New Era for Wind Power in the United States,' US Department of Energy, Washington DC: http://www.energy.gov/windvision

EC (2014) '2030 Framework for Climate and Energy Policies', European Commission, Brussels: http://ec.europa.eu/clima/policies/2030/index_en.htm

Elliott, D. (2015) 'Green Jobs and the Ethics of Energy', in Hersh, M. (ed) 'Ethical Engineering for International Development and

DOI: 10.1057/9781137584434.0008

Environmental Sustainability', Springer, London: http://www.springer.com/gb/book/9781447166177
EP (2012) 'Impact of EU Bioenergy Policy on Developing Countries', European Parliament Briefing paper: http://www.ecologic.eu/files/attachments/Publications/2012/2610_21_bioenergy_lot_21.pdf
ET (2015) 'Clean Energy Push: Drop in Solar Power Cost a Game Changer, Says PM Narendra Modi', *Economic Times*, 16 February: http://economictimes.indiatimes.com/articleshow/46256934.cms
EurObservER (2014) 'Estimates of the Renewable Energy Share in Gross Final Energy Consumption for the Year 2012', ObservER, Paris: http://www.eurobserv-er.org/pdf/press/year_2013/res/english.pdf
GEA (2012) 'Global Energy Assessment', International Institute for Applied Systems Analysis, Austria: http://www.iiasa.ac.at/web/home/research/researchPrograms/Energy/Home-GEA.en.html
Hancock, K. (ed) (2015) 'Renewable Energy in Sub-Saharan Africa', *Energy Research and Social Science*, 5, pp. 1–134, January: http://www.sciencedirect.com/science/journal/22146296/5/
Heinberg, R. (2015) 'Our Renewable Future', Post Carbon Institute, 21 January: http://www.postcarbon.org/our-renewable-future-essay/
Hertwich, E., Gibon, T., Bouman, E., Arvesen, A., Suh, S., Heath, G., Bergesen, J., Ramirez, A., Vega, M. and Shi, L. (2014) 'Integrated Life-Cycle Assessment of Electricity-Supply Scenarios Confirms Global Environmental Benefit of Low-Carbon Technologies', PNAS: http://www.pnas.org/content/early/2014/10/02/1312753111.abstract
Huenteler, J., Anadon, Lee, L. and Santen, N. (2014) 'Commercializing Second Generation Biofuels', Energy Technology Innovation Policy group, Harvard Kennedy School Report: http://belfercenter.ksg.harvard.edu/publication/24889/commercializing_secondgeneration_biofuels.html
Huesemann, M. and Huesemann, J. (2011) 'Technofix: Why Technology Won't Save Us or the Environment', New Society Publishers, Gabriola Island, BC: http://www.newsociety.com/Books/T/Techno-Fix
IEA (2014) 'The Power of Transformation – Wind, Sun and the Economics of Flexible Power Systems', International Energy Agency, Paris: http://www.iea.org/newsroomandevents/pressreleases/2014/february/name,47513,en.html
IPSOS (2011) 'Global Citizen Reaction to the Fukushima Nuclear Plant Disaster', IPSOS Global Advisor, global poll carried out in May 2011:

DOI: 10.1057/9781137584434.0008

http://www.ipsos-mori.com/Assets/Docs/Polls/ipsos-global-advisor-nuclear-power-june-2011.pdf
IRENA (2013) 'Africa's Renewable Future: The Path to Sustainable Growth', International Renewable Energy Agency, Abu Dhabi: http://www.irena.org/menu/index.aspx?mnu=Subcat&PriMenuID=36&CatID=141&SubcatID=276
IRENA (2014a) 'Renewable Energy Prospects: China', REMap 2030 China, International Renewable Energy Agency, Abu Dhabi: http://www.irena.org/remap/IRENA_REmap_China_report_2014.pdf
IRENA (2014b) 'Global Bioenergy Supply and Demand Projections: A Working Paper for REmap 2030', International Renewable Energy Agency, Abu Dhabi: http://www.irena.org/menu/index.aspx?mnu=Subcat&PriMenuID=36&CatID=141&SubcatID=446
IRENA (2015a) 'Renewable Energy Prospects: United States of America', REMap 2030, International Renewable Energy Agency, Abu Dhabi: http://www.irena.org/REmap/IRENA_REmap_USA_report_2015.pdf
IRENA (2015b) 'Renewable Energy and Jobs – Annual Review 2015', International Renewable Energy Agency, Abu Dhabi: http://www.irena.org/menu/index.aspx?mnu=Subcat&PriMenuID=36&CatID=141&SubcatID=585
ISEP (2013) 'Renewables 2013 Japan Status Report,' Institute for Sustainable Energy Policies, Tokyo: http:www.isep.or.jp/en/library/2918
Keller, D., Young, F. and Oschlies, A. (2014) 'Potential Climate Engineering Effectiveness and Side Effects during a High Carbon Dioxide-Emission Scenario', *Nature Communications*, 5, 3304: http://www.nature.com/ncomms/2014/140225/ncomms4304/full/ncomms4304.html
Kroll, M. (2013) 'The Monetary Cost of the Non-Use of Renewable Energies', World Future Council, Hamburg: http://www.climatenewsnetwork.net/2013/01/fossil-fuel-too-valuable-to-burn
Kunze, C. and Becker, S. (2014) 'Energy Democracy in Europe: A Survey and Outlook', Brussels, Rosa-Luxemburg-Stiftung: http://www.rosalux.de/fileadmin/rls_uploads/pdfs/sonst_publikationen/Energy-democracy-in-Europe.pdf
Lahn, G. and Stevens, P. (2011) 'Burning Oil to Keep Cool: The Hidden Energy Crisis in Saudi Arabia', Royal Institute of International

DOI: 10.1057/9781137584434.0008

Affairs, Chatham House, London: http://www.chathamhouse.org/publications/papers/view/180825

Mai, T., Sandor, D., Wiser, R. and Schneider, T. (2012). 'Renewable Electricity Futures Study: Executive Summary', US National Renewable Energy Laboratory, NREL/TP-6A20-52409-ES, Golden, CO: http://www.nrel.gov/docs/fy13osti/52409-ES.pdf

Parkinson, G. (2013) 'RWE Sheds Old Business Model, Embraces New Energy Reality', RENew Economy, 22 October: http://reneweconomy.com.au/2013/rwe-sheds-old-business-model-embraces-new-energy-reality-52967

Parkinson, G. (2014) 'Germany: Decline in Fossil Fuel Generation Is Irreversible', RENew Economy, 5 March: http://reneweconomy.com.au/2014/germany-decline-of-fossil-fuel-generation-is-irreversible-75224

Pelkonen, P., Mustonen, M., Asikainen, A., Egnell, G., Kant, P., Leduc, S. and Pettenella, D. (eds) (2014) 'Forest Bioenergy for Europe', European Forestry Institute: http://www.efi.int/files/attachments/publications/efi_wsctu_4_net.pdf

Sorensen, B. (2014) 'Energy Intermittency', Routledge, London: http://www.routledge.com/books/details/9781466516069/

Steitz, C., and Eckert, V. (2012) 'German Shift from Nuclear a Herculean Task – Execs', quoting RWE CEO Crossman, Reuters, 18 January: http://www.reuters.com/article/2012/01/18/germany-energy-idUSL6E8CI12Y20120118

Teyssen, J. (2012) 'E.ON Confirms Strategy Focused on Renewables', Windpower Monthly, quoting E.ON Executive Teyssen's Comments to German Business Daily Handelsblatt, 30 March: http://www.windpowermonthly.com/go/windalert/article/1124937/?DCMP=EMC-CONWindpowerWeekly

UBA (2014) 'Germany in 2050 – A Greenhouse Gas-Neutral Country', German Federal Environment Agency (UBA), Dessau-Roßlau: http://www.umweltbundesamt.de/publikationen/germany-in-2050-a-greenhouse-gas-neutral-country

WEC (2013) 'World Energy Scenarios: Composing Energy Futures to 2050'. World Energy Council, London: http://www.worldenergy.org/publications/2013/world-energy-scenarios-composing-energy-futures-to-205

DOI: 10.1057/9781137584434.0008

WWF (2013) '100% Renewable Energy by 2050 for India', World Wildlife Fund for Nature, with TERI, New Delhi: http://www.wwfindia.org/?10261/100-Renewable-Energy-by-2050-for-India
WWF (2014) 'China's Future Generation', World Wildlife Fund for Nature: http://worldwildlife.org/publications/china-s-future-generation-assessing-the-maximum-potential-for-renewable-power-sources-in-china-to-2050

DOI: 10.1057/9781137584434.0008

6

Sustainable Futures: What Kind of Mix?

Abstract: *Some hope that the expansion of renewables will allow for continued economic growth while escaping environmental constraints, a classic technical fix. However, there may be limits to economic growth other than energy resources and their impacts, while some think that a move to a more sustainable less growth-oriented economy will be needed. Scenarios based on that essentially decentralist approach, and on more conventional approaches, illustrate the differing implications for energy technology. Some, adopting an economic or technological deterministic view, say we have no real choices, but a rival contention is that there are choices as to the types of society and to the technical mix to sustain them. If the latter is true, we need to make choices soon.*

Keywords: decentralised society; economic growth; green growth; political choices; technological choices

Elliott, David. *Green Energy Futures: A Big Change for the Good.* Basingstoke: Palgrave Macmillan, 2015.
DOI: 10.1057/9781137584434.0009.

6.1 Green growth

A report from the Global Commission on the Economy and Climate, 'Better Growth, Better Climate', says that economic growth and action on climate change can now be achieved together: there are major opportunities in three key sectors of the global economy, cities, land use, energy. By improving efficiency, investing in infrastructure and stimulating innovation across these sectors and the wider economy, governments and businesses can deliver strong growth with lower emissions. The commission's chair, former president of Mexico Felipe Calderón, said the report 'refutes the idea that we must choose between fighting climate change or growing the world's economy. That is a false dilemma,' and 'shows how technological and structural change are driving new opportunities to improve growth, create jobs, boost company profits and spur economic development' (GCEC, 2014).

The United Kingdom's Department of Energy and Climate Change said similar things in 'Securing Our Prosperity through a Global Climate Change Agreement': green growth is possible (DECC, 2014). A core concept underpinning this sort of analysis is that the link between economic growth and growth in energy use, and hence carbon emissions, has been decoupled. Indeed some say that economic growth will produce the surplus and the technology needed to deal with emissions.

While the idea that some countries have learnt how to run, and even expand, on a leaner energy mixture has its attractions, there are methodological problems with this type of assertion. It may be possible for an advanced industrial country like the United Kingdom to reduce emissions while expanding economic activity, for example, in the service sector. However, much of this gain is because it has been able to export some of the dirtier industrial activities to newly developing countries. But it then buys in goods and services from them, while not accounting for their associated emissions. A full carbon account would include them in the national total, focusing on emissions related to total consumption, including imports, rather than just to those from national production, in which case the decoupling might not be so clear (Harrabin, 2015).

Nevertheless, it may be possible to squeeze out all the dirty processes everywhere in time, and clearly the advent of renewable energy technologies offers the opportunity to do that in energy production. Some see renewables as allowing economic growth to continue, free of major finite energy resource constraints and of linked emission impacts. But

DOI: 10.1057/9781137584434.0009

are there limits to how much energy we can obtain from renewables? In terms of resource limits, which might constrain that global project, the most obvious is land. As we have seen, that could limit how much biomass can be used. Wind (increasingly offshore), solar (on rooftops) may not be land-use limited, but there may be material constraints, 'rare earth' minerals especially, although as noted earlier, the resource limits may not be too onerous, and substitutes may be found and recycling practiced. Water is another issue. While fossil and nuclear plants need large amounts of water for cooling, most renewables do not need any for cooling, the main exception being Concentrated Solar Power plants. Certainly water is not something easily available in desert areas where they would be mostly located. But they can also be air-cooled (albeit at lower efficiency) and water can be piped from the sea (at a cost).

The general point is that, although they may add costs, there are, potentially, technical fixes for most of the problems like this, which means growth in energy use is possible, should that be what is wanted. Of course energy waste must be reduced as much as possible. Efficient use of energy use is vital. It saves money and resources and makes it easier for renewables to meet needs. It also reduces any social and environmental impacts from using renewables. They may be small and local, but they should be minimised.

What about other limits to growth? There are absolute limits to the amount of energy that can be extracted from natural energy flows and the sun's energy input to the planet, but these are some way off. The resource is very large. By contrast, if economic, industrial and agricultural activities continue to expand, other limits may become apparent. Land and water are once again obvious issues. To some extent energy can be used to make fresh water (by desalination), but it cannot be used to make land (the odd new island or coastal infill apart), and although we have learnt how to use energy to increase the productivity of land, this can be at the expense of soil quality, run-off pollutants and biodiversity. However, in addition to marginal land, unsuited to farming, there are plenty of desert areas for solar and sea for offshore wind, wave and tidal projects, so for a while absolute limits in terms of land (biomass apart) should not be a major problem.

More generally, there are wider, more general, objections. Not all 'greens' agree with the idea of green growth and are concerned about the continued growth of a consumer society of the current type. Most 'deep greens' seek a shift to a steady state global economy, freed from

DOI: 10.1057/9781137584434.0009

endless pressure to keep expanding so as to support what they see as an unsustainable and undesirable system, with a skewed social structure and a rapacious economy. While renewable energy, increased efficiency and fuel and material substitution can reduce eco-impacts, it is argued that, on a finite planet, there have to be limits on material consumption at some point. Some say that point is already past: the ecological carrying capacity of the planet has already been exceeded, in which case, growth is a problem not a solution. So renewable technologies which simply substitute for conventional technology and enable society to continue unchanged, chasing yet more economic growth, are not helpful and may actually be counterproductive (Dolack, 2015). At best, they are a short-term 'technical fix', and by making it possible to avoid some of the environmental and resource limits on growth based on using fossil fuels, they delay the point when the issue of sustainable consumption has to be faced. It is argued that we all need to rethink how we live, since, quite apart from its environmental impacts, economic growth is not much of a measure of human happiness. We should seek to focus on improving the quality of life – qualitative growth, not quantitative growth, and develop a new ethos of sustainable consumption (Jackson, 2009).

There are problems with this view too. While it has a clear moral dimension, given the very unequal distribution of wealth and power in the world, and the unlikelihood of this being easily changed, for many poor and excluded people, economic growth offers the only realistic hope of moving beyond a minimal subsistence level of existence. For them, growth, if based on the use of renewable energy, not only offers hope of prosperity, but also of a cleaner future, with access to modern energy services that are denied to many at present. So although in the longer term there may be eco-problems with growth, for now that may be the best available option for the poor.

However, this view also has problems. Although the optimistic 'green growth' view is that all can share in growth, the vast inequalities and imbalances that exist at present mean that, as with all economic growth, the benefits may only trickle down slowly and partially to the poorest. There may be valuable short-term social gains, but as the global population grows, and if affluence really does spread, the material limits imposed by the carrying capacity of the planet will still apply and no one will benefit, protected enclaves possibly apart. It is far from clear if that can be avoided even by some of the more extreme 'ecomodernisation' prescriptions, focused on high technology (Brook et al., 2015)

DOI: 10.1057/9781137584434.0009

Ecomodernists see technology-led growth as part of societal progress. However, while growth might be accepted as a way to relieve poverty, for most radical greens it is harder to accept in the case of the affluent in developed countries. There, it is argued, economies have to keep expanding to create market space for rival chunks of capital to continue to compete. That does lock consumers into an endless spiral of consumption, and into what some see as soulless consumer lifestyles (Huesemann and Huesemann, 2011). Clearly 'consumption' has its attractions: the ability to do more interesting and exciting things, aided by clever new technologies and growing wealth, though some say it is more akin to an addiction. Certainly it would be hard to wean most people off consumerism. However, there have been some brave attempts to outline alternative approaches and visions, based on grass roots initiatives, mapping out new ways to live, and the hope that humanity can do better remains strong (Heinberg, 2014; Alexander, 2014).

6.2 How we live

It may be helpful, to further this discussion, to present some simplified polar opposite views. Some see the use of renewable energy as an integral component of a small-scale decentralised form of society, based on more self-sufficient communities, others as just a technical fix allowing the affluent to continue to live much as they already do. In the former case, the type of renewables preferred, to a degree, defines the way of life. Using small-scale solar, biomass and wind, plus possibly small hydro, would probably constrain the economic development of the community, but that is not seen as a problem, since the aim is a low-consumption lifestyle, not based on creating much surplus to sell. In the second version, renewables of all types and scales are just plugged in as substitutes for fossil and perhaps nuclear power, to run an unchanged commercially orientated society, creating surplus to exchange for more extensive and expensive goods and services.

While some ecomodernists retain nuclear as a key element (Brook et al., 2015), in both of these visions, renewables play a role, but the type and scale differ, as does the aim. Both visions, as outlined, are extreme. In reality few people would settle for the bucolic but possibly limited lifestyle implied in the first, and many people living in the second would in fact like to have some elements of the first, that is, less emphasis on

DOI: 10.1057/9781137584434.0009

consumption, more on community. But as stereotypes, they may have some power. Certainly anti-renewables critics sometimes portray renewables as primitive and backward technologies which would only support a dull, subsistence-level existence. Clearly that is not the case. Even in a heavily decentralised version of society, renewables would be able to support a good life. Indeed advocates of decentralisation say it would be a better quality of life, less materialistic and more concerned about community.

That sounds attractive, but there was once a society based on renewables (wind and water power), in the Middle Ages, and only the most romantic would see that social set-up as desirable. However, social structures are not necessarily determined by the technology: rather, those with power choose options that fit their interests. For example, reliance on centralised technologies like nuclear power probably leads to different types of society, and patterns of employment, than would reliance on decentralised technologies.

So there are choices, with an interaction between technology and society (Elliott, 2003, 2015). As Miller et al. put it,

> The key choices involved in energy transitions are not so much between different fuels but between different forms of social, economic, and political arrangements built in combination with new energy technologies. In other words, the challenge is not simply what fuel to use but how to organize a new energy system around that fuel (...) It is not simply a question of whether to build infrastructure for renewable energy systems but rather how to approach such a task and what forms of intertwined social, economic, political, and technological arrangements get built and/or evolve to produce new forms of energy production and consumption. (Miller et al., 2013)

6.3 Future pathways

Moving away from very extreme examples and high-level generalisations, some of the actual ground-level differences in approach, in terms of the type of economy that might be implied and the scale and type of energy technology involved, have been highlighted in recent UK studies. They provide useful examples of choices that may be ahead. The EPSRC-funded *Realising Transition Pathways Research Consortium* of nine UK universities developed three notional transition pathways to 2050:

DOI: 10.1057/9781137584434.0009

- Central Co-ordination: a transition led mainly by the government.
- Market Rules: Led by companies within a broad policy context set by the government.
- Thousand Flowers: Civil society play leads in delivering distributed low carbon energy with community groups and municipal councils playing a major role.

The team's report, 'Distributing Power: A Transition to a Civic Energy Future', focuses on the last pathway. It stands in strong contrast to the first two, in both of which central direction or market direction leads to an emphasis on large-scale technology, including major contributions from nuclear (up to 30 GW), fossil plants with CCS (up to 44 GW) and large wind projects (up to 53 GW, out of a total of up to 80 GW of renewables), and has energy use rising. In Thousand Flowers, the focus instead is on smaller community-scaled supply technology using renewables and more attention to energy saving. Out of the total generating capacity at 2050 of 149 GW, by 2050, 112 GW is renewable based, while annual electricity demand falls to 310 TWh, from 337 TWh. Localised Combined Heat and Power (CHP) using renewable fuels dominates supply, generated from 44 GW of community-scaled biogas-fired CHP and some domestic-scale micro-CHP. Onshore wind comes next at 21 GW, PV is at 16 GW and offshore wind 8.4 GW, and there are smaller hydro, tidal and wave inputs and an even smaller biomass generation input. Unabated coal and gas are phased out entirely by 2030, nuclear declines to 5 GW, about the same as the small coal and gas Carbon Capture and Storage (CCS) elements.

So it is an approach based on local CHP/district heating (DH), with distributed generation providing 50% of final electricity demand, and DH supplying 60% of heat. But it is not a totally decentralised system: there are some large plants still and a new smart grid/DSM systems, and grid upgrades/interlinks would be essential for balancing variable renewables. The report notes that modeling tests show that the proposed system can do that, with no need for extra storage (which it sees as costly and bulky), but some more interconnectors, to add to flexibility. It says that 'high levels of distributed generation in fact make it necessary for higher levels of interconnection at regional, national, and international levels'. In addition 'prosumers (consumers who produce as well as consume electricity) become key actors in this pathway' this 'leading to significant reduction in overall demand'. However, although there

DOI: 10.1057/9781137584434.0009

would be some domestic-level generation, it seems to be that consumer activities would mainly be on the demand side. The report says: 'Price incentives are the main signals used to achieve buy-in to demand side participation in this future,' with citizens being more engaged, 'whether through increased efficiency measures, demand response, or smart metering'.

This seems to be some way from the prosumer supply-side experience in Germany, where many consumers have bought into PV. Perhaps that explains why there is relatively little PV in the Thousand Flowers scenario (a 4.8% overall contribution). The report does accept that there are other possible UK pathways, with more solar and wind. Its biogas emphasis certainly does seem a little optimistic, while much more wind and PV solar could have been included – the potential is large. But the main emphasis in the study is on the transition process and its institutional requirements. The report notes that traditionally, 'UK renewable electricity generation capacity has been built by large-scale commercial developers and/or utilities, whose finances are globally mobile'. The alternative that they look at is based on the proliferation of distributed energy generators, which are owned fully or in part by municipalities, communities or small-scale investors. They say: 'Centralised generation would still be necessary for base-load and peaking capacity. For this to be viable in a distributed generation future where much of our electricity would be generated locally, the government would need to provide the right incentives for new large-scale plant and infrastructure.' However, much would rely on the 'civic' energy sector, defined as 'energy generation by communities, co-operatives, local authorities, town and parish councils or social housing providers', which they say 'currently relies on motivated individuals and communities and often, voluntary work. The development of a decentralised future would require strong project management and professional expertise to deal with a range of technical, financial, legal and administrative issues'.

The Thousand Flowers team opts for a mix of Local Renewable Schemes, Regional Energy Partnerships and Municipally owned Energy Service Companies. The big question is, could they do it? The report says yes, given the right support structures, especially at the municipal level, so as to get the United Kingdom to near zero carbon by 2050 (RTP Engine Room, 2015).

The Thousand Flowers pathway is spelt out in detail (along with the other scenarios) in a working paper (Barton et al., 2014) and has some

DOI: 10.1057/9781137584434.0009

similarities to the 'Patchwork' 2050 decentral pathway explored by the Energy Technologies Institute (ETI) in 'Options Choices Actions: UK Scenarios for a Low Carbon Energy System Transition' (ETI, 2015). That too has local CHP/DH networks playing a significant role, but not to the same degree as in Thousand Flowers. And to compensate, it has more wind (75 GW) and solar (28 GW) and more nuclear (16 GW).

The ETI's parallel 'Clockwork' scenario, in which big companies and government play a central role, is more conventional. It has less renewables (they stay at roughly the level reached by 2020), but more nuclear (40 GW by 2050) and fossil-fueled plants with CCS, though gas is phased out for direct heating. Larger-scale CHP/DH does, however, play a role, co-ordinated by central government. And it has some large biomass with CCS 'carbon negative' plants (BECCS).

The ETI report claims that, under either of their two pathways, 'the UK can achieve an affordable transition to a low carbon energy system over the next 35 years. Our modeling shows abatement costs ranging from 1–2% of GDP by 2050, with potential to achieve the lower end of this range through effective planning'. And it says that 'it is critical to focus resources in the next decade on preparing these options for wide-scale deployment'. It adds: 'By the mid-2020s crucial decisions must be made regarding infrastructure design for the long-term. High levels of intermittent renewables in the power sector and large swings in energy demand can be accommodated at a cost, but this requires a systems level approach to storage technologies, including heat, hydrogen and natural gas in addition to electricity.'

It insists that there should not be a delay:

> It would be a mistake to think the country can wait until efficiency measures have been exhausted before we turn to alternative, low carbon solutions. If the UK waits until the mid-2020s, a lack of supply chain capacity is likely to mean that preferred solutions have to be supplemented by second-choice technologies at far greater expense. In our model, failure to prepare properly leads to a significant escalation in the cost of abatement action by 2050 (to around 3–4% of GDP).

In both ETI Pathways, grid balancing is taken seriously, with demand-side management playing a role, and the ETI is reasonably neutral about the technology options, saying that 'key technology priorities for the UK energy system include: bioenergy, CCS, new nuclear, offshore wind, gaseous systems, efficiency of vehicles and efficiency/heat provision

DOI: 10.1057/9781137584434.0009

for buildings. CCS and bioenergy are especially valuable. The most cost-effective system designs require zero or even "negative" emissions in sectors where decarbonisation is easiest, alleviating pressure in more difficult sectors.'

However, the inclusion of extra nuclear, and its emphasis on large CCS and bioenergy (e.g., large-scale bioconversion), puts their approach, at least in Clockwork, at the 'hard' centralised end of the technology spectrum, and as such it may not be welcome by many greens, who may be more drawn to the Thousand Flowers, although ETI's Patchwork does have its attractions, including plenty of surplus wind-generated electricity for making hydrogen to use to balance the grid when wind is low, although not as much as from the 100 GW of wind capacity in the Pugwash 2050 scenario (Pugwash, 2013). Then again CHP/DH with heat storage could also help with grid balancing, and that is in all the scenarios looked at earlier, although with smaller-scale community units dominating in Thousand Flowers.

Will any of these pathways materialise? Some seem very radical, Thousand Flowers especially, implying a major social as well as technical change, almost socialistic in terms of new patterns of ownership and control. At present the United Kingdom seems to be heading in almost the opposite direction, maybe to ETI's Clockwork. But, as noted, elsewhere different approaches are being adopted, and the debate over which way to go continues.

6.4 Conclusions

As the ETI noted, there is some urgency in moving ahead, and although their specific prescriptions, or the Thousand Flowers version, may not be the definitive answer, it is clear that there are answers to the question posed at the start of this book, essentially, can fossil fuels be phased out? There is much debate on the right mix. Most of the UK scenarios looked at retain some nuclear, some more, some very much less. But earlier, several completely non-nuclear scenarios were mentioned for countries and regions around the world and for the world as whole. The United Kingdom is one of the few countries, at least in the EU, still wedded to large-scale nuclear expansion.

UK government ministers and officials have admitted that it would be possible to do without nuclear, but insist that would make it harder

DOI: 10.1057/9781137584434.0009

to phase out fossil fuel (Randall, 2014). However, it could be counter-argued that it would make it harder and slower to deploy renewables and energy efficiency. It is not as if the United Kingdom does not have the renewable resources. It is probably the best placed in the EU to develop them (Pugwash, 2013; ZCB, 2014). More to the point, there are many different types, available at a range of scales. If technological diversity is important, as many strategists claim, so as to spread risks, then renewables would seem to be a better set of options than nuclear, which, in addition to all its other problems, at present is based on basically one type of technology, which, although allegedly mature, still apparently needs large long-term subsidies.

The pro- and anti-nuclear debate has absorbed much time and energy over the years, to the detriment of what might be seen as a more relevant, interesting and increasingly urgent debate over what sort of renewable/efficiency mix is needed. That has been the focus of this book. The emphasis has been on the technology, on both the supply and demand side, and on the wider issue of what kind of future we want. As has been indicated, there are plenty of conflicts and uncertainties. However, there is no one fixed economically or technologically determined future: there can be choices amongst many possible mixes. There are some environmental constraints, but the biggest constraint is the lack of vision and political commitment. Some specific choices do not have to be made immediately: some options can be left open. But there is an urgent need to decide which way to go in general terms and then to get stuck in to making the changes needed.

Some say the frameworks of governance that exist in most industrialised countries are not up to making changes of the type needed. The current energy regulatory and decision-making systems in countries like the United Kingdom are certainly cumbersome and slow and do not seem able to reflect the changing energy context, the emerging role of new players, local initiatives and the need for long-term rational energy policy development and management. Instead the system often seems to reflect the vested interests of powerful incumbents. In the United Kingdom there have been calls for a new broader, more open 'public value' governance approach (Mitchell et al., 2015).

The wider political framework also may not be conducive to the change of the type needed, with short-term market concerns often dominating. There may be a need not just for better governance, but also for better governments. Some argue that the multi-party systems becoming more

DOI: 10.1057/9781137584434.0009

common in the EU can open up more opportunities for inputting new ideas into government. We will have to wait to see, though not in the United Kingdom for a while.

Elsewhere, attempts to promote bipartisan policymaking, on energy, climate and much else, have not been conspicuously successful, for example, in the heavily polarised United States, and China is ruled by an essentially uncontested unitary elite. We live in challenging, as well as changing, times, with energy and climate policy being central. The technology is available or can be made so, and public support for at least some types of change is strong. Whether our institutions and governments can rise to the challenge remains to be seen.

The prize for rising to the challenge is, however, very significant. Assessments vary, but a report by the German Institute for Climate Policy and Global Sustainability claimed that, if the United States, the EU and China started taking the steps towards using 100% renewable energy by 2050, they could save a combined US$500 billion per year on fuel imports.

In addition, the programme would save the lives of around 1.3 million people who it estimated are killed prematurely by air pollution, while creating 3 million new jobs by 2030, over and above those lost by the phasing out of conventional energy technologies. It would also keep global warming below the 2° C threshold that many scientists believe is the 'point of no return' for climate change, thus avoiding major and potentially rising economic and social costs (Höhne et al., 2015). This may only be a rough interim assessment, but even if it is only partly accurate it gives an idea of the potential benefits available, and, by implication, indicates the need for action.

Hopefully this book will have indicated that there are a range of choices with varying potential social and environmental implications. Most may be obvious and welcome – clean, cheap energy. Others may be less so, or may open up contentious issues. Some concern wide social issues. For example, a draconian approach to energy use, imposed via, for example, punitive energy pricing or tightly controlled carbon rationing permits, might make life hard for some, with the rich buying themselves out of the constraints and black market energy options thriving. We have to be careful to avoid creating a society which is not worth living in. There are also even wider issues. Given the need to reduce global eco-impacts, some see population growth as a key problem. Few would deny the need for modern birth control measures to be made widely available, and an

DOI: 10.1057/9781137584434.0009

optimistic view is that, given equitable social and economic development, population levels should stabilise. However, some say a viable, sustainable, global future will be possible only with a much lower total population, but few would be willing to see tighter limits imposed.

Some other social issues may be less obvious. As has been illustrated, the current boom in PV solar uptake means that consumers are able to exercise some power over energy markets, but the PV cells are sometimes produced by workers who may have poor working conditions and low pay, and little chance to influence markets or technology (Elliott, 2015). A shift to renewables does not automatically change the social relations of production, only of consumption. That said, change is underway in attitudes to technology, and also to fair trade, at all levels, and that may include changes in working relations and conditions, as part of what has been called a 'Just Transition' (ITUC, 2009).

At the same time, the wider process of change that is underway, and the clear progress being made with renewables, should provide a basis for a response to the rhetorical question sometimes raised: 'if renewables are so wonderful, why aren't they taking off?'. The simple answer is, they are, as I hope this book has shown. Moreover they can expand to meet all our needs, given the right support.

6.5 References

Alexander, G. (2014) 'eGaia: Growing a Peaceful, Sustainable Earth through Communications', Earth Connected E-book: http://earthconnected.net/egaia-2nd-edition/buy-or-download-egaia/

Barton, J., Davies, L., Foxon, T., Galloway, S., Hammond, G., O'Grady, A., Robertson, E. and Thomson, M. (2013) 'Transition Pathways for a UK Low Carbon Electricity System: Comparing Scenarios and Technology Implications', Realising Transition Pathways Working Paper, December: http://www.realisingtransitionpathways.org.uk/realisingtransitionpathways/publications/WorkingPapers.html

Brook, B. et al. (2015) 'An Ecomodernist Manifesto', produced by 18 academics, April: http://www.ecomodernism.org/manifesto/

DECC (2014) 'Securing Our Prosperity through a Global Climate Change Agreement', UK Department of Energy and Climate Change: http://www.gov.uk/government/news/prosperity-and-growth-hand-in-hand-with-carbon-reduction

DOI: 10.1057/9781137584434.0009

Dolack, P. (2015) 'Renewable Energy Alone Cannot Reverse Global Warming or Make a Sustainable World', *The Ecologist*, 25 May: http://www.theecologist.org/News/news_analysis/2861727/renewable_energy_alone_cannot_reverse_global_warming_or_make_a_sustainable_world.html

Elliott, D. (2003) 'Energy, Society and Environment', Routledge, London: http://www.routledge.com/books/details/9780415304863/

Elliott, D. (2015) 'Green Jobs and the Ethics of Energy', in Hersh, M. (ed) 'Ethical Engineering for International Development and Environmental Sustainability', Springer, London: http://www.springer.com/gb/book/9781447166177

ETI (2015) 'UK Scenarios for a Low Carbon Energy System Transition: Options, Choices, Action', Energy Technologies Institute, Loughborough and Birmingham: http://theeti.cmail2.com/t/ViewEmail/j/3815A8D22B622CAE/2558A135EF6BFBF74D402EFBD42943A3

GCEC (2014), 'Better Growth, Better Climate', Global Commission on the Economy and Climate, Special Report: http://newclimateeconomy.report/

Harrabin, R. (2015) 'CO_2 Cuts Claim Sees Ministers Challenged by Experts', BBC News Item, 19 March: http://www.bbc.co.uk/news/science-environment-31952888

Heinberg, R. (2014) 'How to Shrink the Economy without Crashing It: A Ten-Point Plan', Resilience/Post Carbon Institute: http://www.resilience.org/stories/2014-11-04/how-to-shrink-the-economy-without-crashing-it-a-ten-point-plan

Höhne, N., Day, T., Hänsel, G. and Fekete, H. (2015) 'Assessing the Missed Benefits of Countries' National Contributions', New Climate – Institute for Climate Policy and Global Sustainability gGmbH, Berlin, Report for the Climate Action Network: http://www.sciencealert.com/switching-to-100-renewables-by-2050-would-save-major-economies-500-billion

Huesemann, M. and Huesemann, J. (2011) 'Technofix: Why Technology Won't Save Us or the Environment', New Society Publishers, Gabriola Island, BC: http://www.newsociety.com/Books/T/Techno-Fix

ITUC (2009) 'What Is Just Transition?' International Trade Union Confederation, Brussels: http://www.ituc-csi.org/what-s-just-transition

Jackson, T. (2009) 'Prosperity without Growth', Sustainable Development Commission Report, then an Earthscan Book, London: http://www.sd-commission.org.uk/publications.php?id=914

DOI: 10.1057/9781137584434.0009

Miller, C., Iles, A. and Jones, C. (2013) 'The Social Dimensions of Energy Transitions', *Science as Culture Forum on Energy Transitions, Science as Culture*, 22 (2): http://www.tandfonline.com/toc/csac20/22/2#.VSp3DIX1uv8
Mitchell, C., Woodman, B., Kuzemko, C. and Hoggett, R. (2015) 'Working Paper: Public Value Energy Governance', University of Exeter Energy Policy Group, 20 March: http://projects.exeter.ac.uk/igov/working-paper-public-value-energy-governance/1
Pugwash (2013) 'Pathways to 2050: Three Possible UK Energy Strategies,' British Pugwash Group, London: http://www.britishpugwash.org/recent_pubs.htm
Randall, K. (2014) Katherine Randall, 2050 Team Leader at the Department for Energy and Climate Change, in evidence to the House of Lords Select Committee on Science and Technology review of Nuclear Research and Development Capabilities, para 40, 3rd Report of Session 2010–2012, November: http://www.publications.parliament.uk/pa/ld201012/ldselect/ldsctech/221/221.pdf
RTP Engine Room (2015) 'Distributing Power: A Transition to a Civic Energy Future', Report of the Realising Transition Pathways Research Consortium 'Engine Room': http://www.realisingtransitionpathways.org.uk/realisingtransitionpathways/news/distributing_power.html
ZCB (2014) 'Zero Carbon Britain', Centre for Alternative Technology Report: http://www.zerocarbonbritain.com/index.php/zcb-latest-report

DOI: 10.1057/9781137584434.0009

7
Afterword

Elliott, David. *Green Energy Futures: A Big Change for the Good*. Basingstoke: Palgrave Macmillan, 2015.
DOI: 10.1057/9781137584434.0010.

DOI: 10.1057/9781137584434.0010

Inevitably, this book does not cover all aspects of the many large issues it raises. For example, it hardly touches on transport issues, although some non-fossil transport fuel source options were mentioned. It is an area where change is probably hardest and technical fixes (e.g., using biogas, hydrogen or electricity) may be the most people in the industrialised countries are prepared to consider. That may have to change, with more emphasis on public transport, walking and cycling. To make that viable, there will be a need for new approaches to urban planning, housing and retailing as well as employment and leisure location patterns. A big topic. The same goes for agriculture and food production. That, and our dietary choices, have huge energy and climate implications. For example, can we continue to use energy (and fertilizers) to increase the productivity of land? But that is a topic for another book!

So is the wider debate on population, touched on briefly earlier. For the moment, the poor actually consume very little compared to the affluent minority, though that may change if consumption patterns of the current type spread. In which case a key issue is whether an equitable sustainable economy and society can be developed that avoids the need for draconian population controls. Again a big topic, for another book.

As far as the present book is concerned, the emphasis has been on the technological options in energy supply, and if you need more information and analysis on that, there is a further reading and web link guide in the following page, as well as the extensive end-of-chapter references in this text, with web links, all viewed and live in July 2015.

Hopefully, this book will have conveyed both the enthusiasm for change that abounds in the renewable energy community and the sense of urgency that follows when the scale of the environmental problems are understood. As an aid to that, I would like to recommend this fascinating animation showing the growing historical pace of carbon dioxide additions to the atmosphere: http://www.youtube.com/watch?v=vA7tfz3k_9A

However, as a counter to over-optimism about what technology can do in response, I recommend the admittedly rather bleak 'Critique of Techno-Optimism' by Samuel Alexander, produced in 2014 for the Australian Post Carbon Pathways project: http://www.postcarbonpathways.net.au/2014/01/23/working-paper-series-1st-edition/#.VZvyEIX1uv8. Some of the project's other output is more positive!

DOI: 10.1057/9781137584434.0010

Further reading and updates

As a short guide to a wider area, this book crams in a lot of material and you may wish to follow some of the details up, or get other perspectives. For a very broad view see Walt Patterson's 'Electricity v Fire': http://www.waltpatterson.org/evf.htm.

For more technical details there are some good standard text books on renewables:

Harvey, D. (2010) 'Carbon Free Energy' Earthscan.

Boyle G. (ed) (2012) 'Renewable Energy', Oxford University Press (3rd edition).

Twiddel, J. and Weir, T. (2015) 'Renewable Energy Resource', Routledge (3rd edition)

My 2013 e-book *Renewables: A Review of Sustainable Energy Supply Options* for the Institute of Physics also takes you through the basics focusing on the technology. It is updated *weekly* by my *Renew Your Energy* Blog for the IOP's Environmental Research Web: http://blog.environmentalresearchweb.org/author/dxe

My bimonthly *Renew on Line* Blog, from which much of the material for this book was taken, may also be useful: https://renewnatta.wordpress.com/. For a more scholarly text see the book I edited, with my new update: *Sustainable Energy* (2015 edition), Palgrave.

DOI: 10.1057/9781137584434.0010

Index

DOI: 10.1057/9781137584434.0011

DOI: 10.1057/9781137584434.0011

DOI: 10.1057/9781137584434.0011

DOI: 10.1057/9781137584434.0011